# Communications
# in Computer and Information Science     1183

*Commenced Publication in 2007*
Founding and Former Series Editors:
Simone Diniz Junqueira Barbosa, Phoebe Chen, Alfredo Cuzzocrea,
Xiaoyong Du, Orhun Kara, Ting Liu, Krishna M. Sivalingam,
Dominik Ślęzak, Takashi Washio, Xiaokang Yang, and Junsong Yuan

More information about this series at http://www.springer.com/series/7899

Abbas M. Al-Bakry · Safaa O. Al-Mamory ·
Mouayad A. Sahib · Haitham S. Hasan ·
George S. Oreku · Thaker M. Nayl ·
Jaafar A. Al-Dhaibani (Eds.)

# New Trends in Information and Communications Technology Applications

4th International Conference, NTICT 2020
Baghdad, Iraq, June 15, 2020
Proceedings

 Springer

*Editors*
Abbas M. Al-Bakry
University of Information Technology
and Communications
Baghdad, Iraq

Mouayad A. Sahib
University of Information Technology
and Communications
Baghdad, Iraq

George S. Oreku
Open University of Tanzania
Dar es Salaam, Tanzania

Jaafar A. Al-Dhaibani
University of Information Technology
and Communications
Baghdad, Iraq

Safaa O. Al-Mamory ⓘ
University of Information Technology
and Communications
Baghdad, Iraq

Haitham S. Hasan
University of Information Technology
and Communications
Baghdad, Iraq

Thaker M. Nayl
University of Information Technology
and Communications
Baghdad, Iraq

ISSN 1865-0929            ISSN 1865-0937 (electronic)
Communications in Computer and Information Science
ISBN 978-3-030-55339-5            ISBN 978-3-030-55340-1 (eBook)
https://doi.org/10.1007/978-3-030-55340-1

This Springer imprint is published by the registered company Springer Nature Switzerland AG
The registered company address is: Gewerbestrasse 11, 6330 Cham, Switzerland

# Preface

The 4th International Conference on New Trends in Information and Communications Technology Applications (NTICT 2020), was hosted and organized by the University of Information Technology and Communications, Iraq. The NTICT conference was planned to take place in Baghdad during March 11–12, 2019, however, due to the COVID-19 pandemic the conference was postponed to June 15, 2020, and organized virtually. NTICT is an international conference focusing on specific topics in computer networks and machine learning. NTICT 2018 was the first conference in the series to have its proceedings be published in *Communications in Computer and Information Science* (CCIS) by Springer. NTICT 2020 aimed at providing a meeting for an advanced discussion of evolving applications in machine learning and computer networks. The conference brought together both young researchers and senior experts to share novel findings and practical experiences in the aforementioned fields.

The NTICT 2020 conference enhanced the scientific research on this subject in Iraq and the rest of the world. The call for papers resulted in a total of 90 submissions from around the world. Every submission was assigned to at least three members of the Program Committee for review. The Program Committee decided to accept only 18 papers based on a highly selective review process, which were arranged into three sessions, resulting in a strong program with an acceptance rate of about 20%. The accepted papers were distributed in two main tracks: Networks, and Machine Learning Approaches. We would like to thank all who contributed to the success of this conference, in particular the members of the Program Committee (and the additional reviewers) for carefully reviewing the contributions and selecting a high-quality program. The reviewer's efforts to submit the review reports within the specified period were greatly appreciated. Their comments were valuable and very helpful in the selection process. Furthermore, we would like to convey our gratitude to the keynote speakers for their excellent presentations. The words of thanks is extended to all authors who submitted their papers and for letting us evaluate their work. The submitted papers were managed using the Open Conference System (OCS); thanks to the Springer team for making these proceedings possible. We hope that all participants enjoyed a successful conference, made a lot of new contacts, and engaged in fruitful discussions.

June 2020

Abbas M. Al-Bakry
Safaa O. AL-Mamory
Mouayad A. Sahib
Haitham S. Hasan
George S. Oreku
Thaker M. Nayl
Jaafar A. Al-Dhaibani

# Organization

## General Chair

Abbas M. AL-Bakry — President of University of Information Technology and Communications, Iraq

## Co-chair

Hasan Shaker Magdy — Dean of Al-Mustaqbal University College, Iraq

## Organizing Committee

Abbas M. Al-Bakry — University of Information Technology and Communications, Iraq

Safaa O. AL-Mamory — University of Information Technology and Communications, Iraq

Mouayad A. Sahib — University of Information Technology and Communications, Iraq

Haitham S. Hasan — University of Information Technology and Communications, Iraq

George S. Oreku — Open University of Tanzania, Tanzania, and North West University South Africa, South Africa

Thaker M. Nayl — University of Information Technology and Communications, Iraq

Jaafar A. Al-Dhaibani — University of Information Technology and Communications, Iraq

## Steering Committee

Jane J. Stephan — University of Information Technology and Communications, Iraq

Sinan A. Naji — University of Information Technology and Communications, Iraq

Zaidon A. Abdulkariem — University of Information Technology and Communications, Iraq

Ahmed A. Hashim — University of Information Technology and Communications, Iraq

Ali Hassan Tarish — University of Information Technology and Communications, Iraq

Ahmed Sabah Ahmed — University of Information Technology and Communications, Iraq

| Inaam Rikan Hassan | University of Information Technology and Communications, Iraq |
| Hussanin Yaarub Mohammed | Al-Mustaqbal University College, Iraq |
| Ali Hussein Shaman | Al-Mustaqbal University College, Iraq |
| Samir M. Hameed | University of Information Technology and Communications, Iraq |

## International Scientific Committee

| Abdel-Badeeh Salem | Ain Shames University, Egypt |
| Amir Masoud Rahmani | Islamic Azad University, Iran |
| Jane J. Stephan | University of Information Technology and Communications, Iraq |
| Buthaina Fahran Abd | University of Information Technology and Communications, Iraq |
| Alaa Hussein al-Hammami | Princess Sumaya Univesrsity for Science and Technology, Jordan |
| Alaa Al-Shammery | University of Information Technology and Communications, Iraq |
| Ali Abdulhadi Jasim | University of Information Technology and Communications, Iraq |
| Hamza Mohammed Ridha | Al-Mustaqbal University College, Iraq |
| Najeh Rasheed Jooda | Al-Mustaqbal University College, Iraq |
| Ali Al-Sherbaz | University of Northampton, UK |
| Ali H. Al-Timemy | University of Baghdad, Iraq |
| Dennis Lupiana | Institute of Finance Management, Tanzania |
| Eva Volna | University of Ostrava, Czech Republic |
| George Oreku | Open University of Tanzania, Tanzania, and North West University South Africa, South Africa |
| Hasan Fleyeh | Dalarna University, Sweden |
| Hilal Mohammed | Applied Science University, Bahrain |
| Hoshang Kolivand | Liverpool John Moores University, UK |
| Mahdi Nsaif Jasim | University of Information Technology and Communications, Iraq |
| Malleswara Talla | Concordia University, Canada |
| Mohammad Shojafar | University Sapienza of Rome, Italy |
| Mohanad Alhabo | University of Leeds, UK |
| Mouayed Sahib | University of Information Technology and Communications, Iraq |
| Mudafar Kadhim Ati | Abu Dhabi University, UAE |
| Omar Daoud | Philadelphia University, Jordan |
| Osama A.H. Al-Tameemi | University of Central Florida, USA |
| Safaa O. Al-Mamory | University of Information Technology and Communications, Iraq |

| | |
|---|---|
| Hussein A. Mohammed | University of Information Technology and Communications, Iraq |
| Huda Kadim Tayh | Università of Information Technology and Communications, Iraq |
| Sattar J. Al-Saedi | University of Information Technology and Communications, Iraq |
| Yaser Alfadhl | Queen Mary University of London, UK |
| Zahir Hussein | Al-Kufa University, Iraq |
| Nabeel H. Kaghed | Ministry of Higher Education and Scientific Research, Iraq |
| Eman Alshamery | University of Babylon, Iraq |
| Nael Ahmed Al-Shareefi | University of Information Technology and Communications, Iraq |
| Saif Q. Mohammed | University of Information Technology and Communications, Iraq |
| Hassanain M. Hassan | University of Information Technology and Communications, Iraq |
| Ghaidaa Al-Sultany | University of Babylon, Iraq |
| Abbas M. Ali | Salahaddin University, Iraq |
| Ahmed Abdulsahib | University of Information Technology and Communications, Iraq |
| Ahmed Al-Azawei | University of Babylon, Iraq |
| Ahmed AL-Jumaili | University of Information Technology and Communications, Iraq |
| Aini Syuhada Md Zain | Universiti Malaysia Perlis, Malaysia |
| Alejandro Zunino | ISISTAN-CONICET, Argentina |
| Ali Kadhum Idrees | University of Babylon, Iraq |
| Ali Obied | University of Sumer, Iraq |
| Ana Paula Lopes | Polytechnic of Porto, Portugal |
| Ashery Mbilinyi | University of Basel, Switzerland |
| Boxiang Dong | Montclair State University, USA |
| Ching-Lueh Chang | Yuan Ze University, Taiwan |
| Christian Esposito | University of Napoli Federico II, Italy |
| Daniel Hunyadi | Lucian Blaga University of Sibiu, Romania |
| Dhiah Al-Shammary | University of Al-Qadisiyah, Iraq |
| Mehdi Ebady Manaa | University of Babylon, Iraq |
| Eduardo Fernandez | Florida Atlantic University, USA |
| Fazidah Othman | University of Malaya, Malaysia |
| Francesco Colace | Università degli Studi di Salerno, Italy |
| Ghina Saad | American University of Culture and Education, Lebanon |
| Haider K. Hoomod | Al Mustansiriyah University, Iraq |
| Haitham Badi | University of Information Technology and Communications, Iraq |
| Hamidah Ibrahim | Universiti Putra Malaysia, Malaysia |
| Hasan Al-Khaffaf | University of Duhok, Iraq |

Hassan Harb                    American University of Culture and Education,
                               Lebanon
Hayder Safi                    Altinbaş University Istanbul, Turkey
Hilal Adnan Fadhil             Al-Farabi University, Iraq
Hoshang Kolivand               Liverpool John Moores University, UK
Jaafar A. Al-Dhaibani          University of Information Technology
                               and Communications, Iraq
Jose Angel Banares             University of Zaragoza, Spain
Kuldeep Kumar                  National Institute of Technology Jalandhar, India
Luca Cagliero                  Politecnico di Torino, Italy
M. Ali Aydin                   İstanbul University, Turkey
Mahdi Abed Salman              University of Babylon, Iraq
Marco Brocanelli               Wayne State University, USA
Marwa Ibrahim                  American University of Culture and Education,
                               Lebanon
Matobola Mihale                Open University of Tanzania, Tanzania
Miguel Carriegos               RIASC, Spain
Mohamad Abou Taam              American University of Culture and Education,
                               Lebanon
Mohamad Nizam Ayub             University of Malaya, Malaysia
Mohammad Abdel-majeed          University of Jordan, Jordan
Mohammad Rasheed               University of Information Technology
                               and Communications, Iraq
Mohammed Al-Shuraifi           University of Babylon, Iraq
Mudhafar Ali                   Al-Iraqia University, Iraq
Mushtaq Alqiasy                Al-Iraqia University, Iraq
Nadia Essoussi                 Higher Institute of Management, Tunisia
Nornazlita Hussin              University of Malaya, Malaysia
Oleksandr Lemeshko             Kharkiv National University of Radio Electronics,
                               Ukraine
Oleksandra Yeremenko           Kharkiv National University of Radio Electronics,
                               Ukraine
Ovidiu Stefan                  Technical University of Cluj Napoca, Romania
Rashid Ali Fayadh              Middle Technical University, Iraq
Razali Ngah                    Universiti Teknologi Malaysia, Malaysia
Rosilah Hassan                 Universiti Kebangsaan Malaysia, Malaysia
Safanah Raafat                 University of Technology, Iraq
Samaresh Bera                  IIT Kharagpur, India
Santhosh Kumar                 GMR Institute of Technology, India
Shuai Liu                      Inner Mongolia University, China
Sinan Naji                     University of Information Technology
                               and Communications, Iraq
Somkait Udomhunsakul           Rajamangala University of Technology Suvarnabhumi,
                               Thailand
Suhad Ahmed Ali                University of Babylon, Iraq
Sura Alrashid                  University of Babylon, Iraq

Suresh P.                        Vel Tech Rangarajan Sagunthala R & D Institute
                                 of Science and Technology, India
Tariq Fadil                      Dijla University College, Iraq
Thaker M. Nayl                   University of Information Technology
                                 and Communications, Iraq
Wesam Bhaya                      University of Babylon, Iraq
Yaseen Jurn                      University of Information Technology
                                 and Communications, Iraq
Zaid Ali                         Al-Mansour University College, Iraq

## Secretary Committee

Jwan Knaan Alwan                 University of Information Technology
                                 and Communications, Iraq
Alya Jemma                       University of Information Technology
                                 and Communications, Iraq
Ali Abed Al-Kareem               University of Information Technology
                                 and Communications, Iraq
Ali Dakhel Hussan                University of Information Technology
                                 and Communications, Iraq
Ammar AbdRaba Sakran             University of Information Technology
                                 and Communications, Iraq
Hiba Muhmood Yousif              University of Information Technology
                                 and Communications, Iraq

# Contents

# Computer Networks

# Latency Evaluation of an SDN Controlled by Flowvisor and Two Heterogeneous Controllers

Zainab AL-Badrany[✉] and Mohammed Basheer Al-Somaidai[✉]

University of Mosul, Mosul, Iraq
zainabalbadrany@gmail.com,
mohammedbasheerabdullah@uomosul.edu.iq

**Abstract.** Network slicing and SDN are essential requirements of implementing 5G technology in networks. Virtualization is achieved through hypervisors. Flowvisor is one of SDN hypervisors, it establishes multiple virtual SDN networks (vSDNs) to allow users to share the same network in controlled and isolated manner. Different services, such as voice and video, can run on isolated virtual slices to improve the quality of service across applications and to optimize the network performance. This paper presents hierarchical model of SDN controllers with Flowvisor as slicing tool and two isolated heterogeneous controllers, POX and Ryu. The aim of this paper is to recognize network latency for each slice. The results revealed an improvement in reducing the round trip time by 31% when both controllers (Ryu and Pox) have the same link bandwidth, which makes the Ryu controller more suitable than POX for real time applications. Also the results confirm that as the number of shared switches between the two slices increase, the delay factor becomes more effective.

**Keywords:** Software Defined Network · Network slicing · Latency · Mininet · Flowvisor · Ryu · POX · D-ITG tool

## 1 Introduction

Generally, the increasing demand for higher communication speed, reliability and scalability has created a need for better network control and management [1]. Software Defined Network (SDN) seems to be the best solution through its new concept separating the control plane and the data plane. This segregation enhances flexibility, programmability, cost effectiveness, and dynamically management through the SDN controller. OpenFlow protocol is the dominant standard communication protocol between the controller and the switches [2]. The infrastructure of the 5G networks is based on SDN to provide open communications system and to handle applications dynamically [3]. SDN achieves the required QoS and provides Network as a Service (NaaS) through network slicing. Network slicing uses network virtualization to build multiple virtual

Thanks and gratitude to Mosul University for providing facilities.

networks (slices) on shared infrastructure [4]. A slice can be defined to an application independent of other applications slices. The past years have witnessed a lot of work towards using network virtualization in the SDN environment [5]. Network virtualization is an important topic in communication networks, for its benefits in isolation and enabling multiple applications to access the network resources. Network virtualization refers to an abstracting layer that permits the flexible sharing and slicing of network resources among multiple virtual networks. Each user executes its applications over its virtual network (slice of the actual physical network) independent of the other users [6]. The virtualization of SDN networks promises new innovative opportunities due to the combined benefits of SDN networking and network virtualization. Combining SDN and Network Virtualization (NV) introduces network programmability, flexibility, scalability, automation, shared access, topology isolation, and simplifies management with lower cost [7].

An SDN network with the concept of SDN hypervisors should consist of one or more SDN controllers to manage the entire network and to allow multiple network operation systems accessing the same network [8]. Virtualizing SDN networks through an SDN hypervisor could abstract the physical SDN network into multiple logically isolated virtual SDN networks (vSDNs). Each vSDN "slice of the network" can operate its own virtual SDN network using its own controller, independent of the other slices through the OpenFlow protocol [9].

Several solutions have been proposed to create network slices. Rexford et al. [6] proposed virtual local area networks (VLANs), that create slices at the data link layer. Salvadori et al. designed AdVisor [10], that creates slices of the entire network to improve abstraction mechanism that hides physical switches in virtual topologies. In order to achieve a more flexible topology abstraction, Corin et al. designed VeRTIGO [11], which is an extension of AdVisor that allows the vSDN controllers to select the desired level of virtual network abstraction and provides the entire set of assigned virtual resources, with full virtual network control and flexibility. Also, Jin et al. presented CoVisor [12], which is built on the concept of the compositional hypervisor and focuses on the cooperation of heterogeneous SDN controllers (controllers written in different programming Languages), on the same network traffic. It focuses on improving the performance of the physical SDN network and it provides mechanisms to protect the physical SDN network against malicious controllers.

## 2  Flowvisor

FlowVisor is an open source, Java-based tool for network slicing, where an SDN can have distributed controllers to control same sets of switches [13]. It is an SDN hypervisor that virtualizes the network, to enable multiple controllers to coexist and share the control for the same physical infrastructure [14]. It acts as intermediate layer between the controllers and the switches, to ensure that a specific controller can only control the switches it is supposed to [15]. Figure 1 shows an abstract view of a virtual SDN.

FlowVisor that was developed at Stanford University uses the OpenFlow protocol to control the underlying network. Flow space is a set of flows running on a topology of switches, which is used for slicing the network into different slices and enforcing

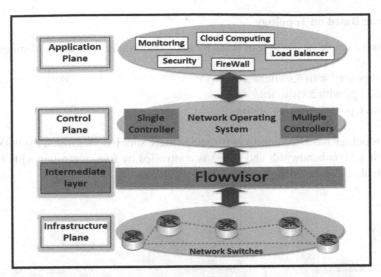

**Fig. 1.** Abstract view of virtual SDN.

strong isolation between them [16]. This isolation means that the data traffic in one slice cannot be captured by hosts in the other slice and that if one slice becomes malicious then it doesn't impact other slices. Each slice should have only single controller. This slice consists of several flow spaces. Each flow space has a rule, much like a firewall, where the actions can be either allow, deny or read only. These actions restrict or permit the controllers to interact with each other, and can also allow one or more switches to belong to two or more slices. FlowVisor slices the network along multiple dimensions, including topology, bandwidth, and forwarding tables in each switch [17, 18].

Slices can be defined to any appropriate application independent of other slices. The delay sensitive applications such as VoIP, video streaming, and other real-time applications, need to be isolated from simple traditional applications to optimize quality and reliability of average transmission rate [8, 9]. Flowvisor Slicing Strategy is the solution to deliver different Quality of Service (QoS) in an optimized manner, by adjusting bandwidth dynamically for each application based on QoS required [7, 8].

## 3  Methodology

A PC with Intel core$^{TM}$–i5 3320 MB CPU @2.60 GHz x4 with RAM 8 GB, 128 SSD, 500 GB hard disk with Ubuntu 14.04 Lts, as Operating System were used. We built an SDN network using Mininet emulator (ver. 2.3.0) with Open vSwitch, and the network was controlled by POX [19] and Ryu (ver. 4.21) [20] as controllers, with Flowvisor (ver.1.4) [21] as SDN hypervisor. The controllers and the switches support OpenFlow protocol (ver 1.0).

## 3.1 Slicing Based on Topology

The following topologies with four hosts for each one, where built using Mininet.

- Linear topology with 4 switches.
- Star topology with 5 switches.
- Diamond topology with 4 switches.

Each topology has sliced up to two slices, where slice1 is controlled by POX controller with 1 Mb/s bandwidth and slice2 is controlled by Ryu controller with 1 Mb/s bandwidth also. Figure 2 demonstrates these topologies:-

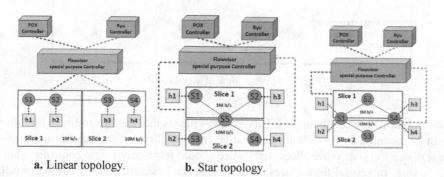

**a.** Linear topology.          **b.** Star topology.

**Fig. 2.** Types of topologies.

## 3.2 Choosing Suitable Controller Based on Bandwidth

Three cases of different bandwidths were taken for the diamond topology that is shown in Fig. 2c. POX controller was used to control the data transfer traffic, while Ryu controller was used to control the voice and video traffic.

- POX controller with 1 Mb/s bandwidth to control slice1, and Ryu controller with 10 Mb/s bandwidth to control slice2.
- Both controllers with 1 Mb/s bandwidth.
- POX controller with 10 Mb/s bandwidth to control slice1, and Ryu controller with 1 Mb/s bandwidth to control slice2.

A flowchart of the methodology that was used to measure the delay time for the controllers is shown in Fig. 3. This delay is needed to gather control and state information from the switches through the flowvisor. Two test tools were used to measure the network performance parameters according to the abilities of these tools. The Mininet test tool "Ping" command was used to measure the end-to-end Round Trip Time (RTT). The other test tool is the D-ITG test tool, which was carried out to measure the average delay and the average jitter. This test tool was used with the following assumptions:

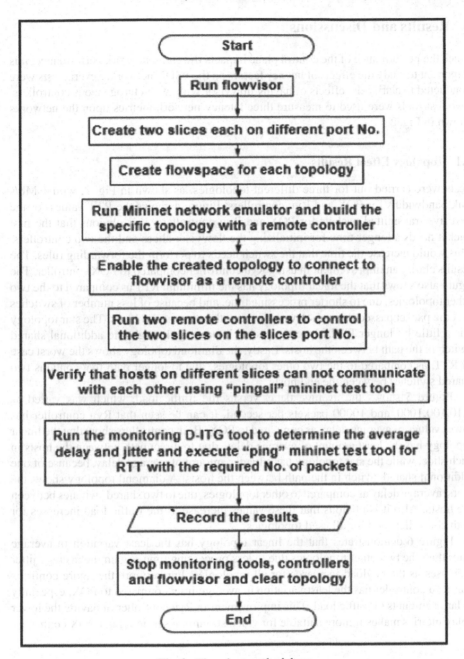

**Fig. 3.** Flowchart methodology.

- Packet size is 512 byte.
- Test duration is 60 s.
- The payload parameter (No. of packets per second) was changed exponentially from one to ten thousands in step of ten to the power one.
- Each topology has four hosts.

# 4   Results and Discussions

Since the performance of the control plane impacts the entire network performance, it is important to study the effects of latency factors on the vSDN network. Several tests were conducted to clarify the effects of having flowvisor with two heterogeneous controllers. Two test tools were used to measure three latency network metrics upon the networks shown in Fig. 2.

## 4.1   Topology Effect Results

Tests were carried out for three different topologies, as shown in Fig. 2, with 1 Mb/s link bandwidths for POX and Ryu controllers. Figure 4 shows the RTT values for the first five transmitted packets for POX and Ryu controllers. It is obvious that the first packet needs a longer time for initializing the flowvisor slices and the two controllers. This would increase the time that the switch needs to perform the forwarding rules. The results clarify an improvement in using Ryu controller compared to POX controller. The figure also shows that the linear topology has the minimum RTT as compared to the two other topologies, due to shorter processing time, and because of less number of switches that the packet passed through traversing the path between end hosts. The star topology has a little bit longer RTT, since the packet is passing through one additional shared switch in the path between the hosts. Lastly, the diamond topology shows the worst case of RTT, as compared to the two other topologies, due to longer path that contains two shared switches between the end hosts.

Figure 5 shows the average delay versus the traffic load, which was varied as 1,10,100,1000, and 10000 packets per second. It can be seen that Ryu controller has lower values of average delay as compared to POX. Both controllers show that the linear topology has the least amount of average delay, due to direct path between the hosts in each slice, while the star topology has a longer amount of average delay, because of one additional shared switch in the path between the hosts. A diamond topology shows the worst average delay as compared to other topologies, due to two shared switches between the hosts. Also it is obvious that these values increase as the traffic load increases for both controllers and for all used topologies.

Figure 6 demonstrates that the linear topology has the least variation in average jitter than the two other topologies due to less processing time. Again the average jitter increases as the payload increases for all topologies. Furthermore the figure confirms that Ryu controller has the least variation in average jitter compared to POX, especially in large amounts of traffic load. This improvement in Ryu controller of having the lower delay metrics makes it more suitable for video streaming services than POX controller.

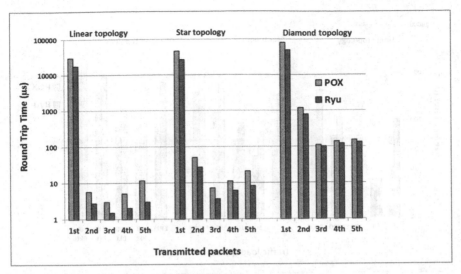

**Fig. 4.** RTT for the three topologies.

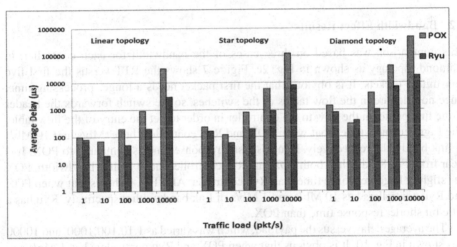

**Fig. 5.** Average delay for the three topologies.

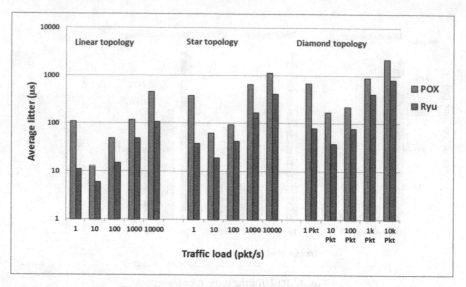

**Fig. 6.** Average jitter for the three topologies.

## 4.2 Bandwidth Effect Results

This comparison was based on three cases of the bandwidth for each controller, in diamond topology as shown in Fig. 2c. Figure 7 shows the RTT versus the first five transmitted packets. It is obvious that the first packet needs a longer processing time, since no rules are in the flow tables of the switches, so the switch forwards the header of the first packet in the flow to the controller in order to get the entry of the flow table. The Figure demonstrates that when POX and Ryu controllers have 1Mb/s and 10 Mb/s of link bandwidth respectively, Ryu has faster response time as compared to POX. It is clear from Fig. 8 that when both POX and Ryu controllers have 1 Mb/s bandwidth, POX has slightly larger response time than Ryu controller. Also Fig. 9 shows that when POX and Ryu controllers has 10 Mb/s and 1 Mb/s of link bandwidth respectively, Ryu has a little bit shorter response time than POX.

The average delay versus the payload, which was varied as 1,10,100,1000, and 10000 was shown in Fig. 10. It is obvious that when POX and Ryu controllers has 1 Mb/s and 10 Mb/s of link bandwidth respectively, Ryu has least average delay as compared to POX. It is obtained from Fig. 11 that when both POX and Ryu controllers have 1 Mb/s bandwidth, POX has an extra average delay over Ryu controller. Only when POX and Ryu controllers have 10 Mb/s and 1 Mb/s bandwidth respectively, POX has a little bit lower average delay than Ryu as shown in Fig. 12.

Lastly, Fig. 13 shows that when POX and Ryu controllers have 1 Mb/s and 10 Mb/s of link bandwidth respectively, Ryu has the least variation in average jitter as compared to POX. It can be noticed from Fig. 14 that when both POX and Ryu controllers have 1 Mb/s bandwidth, POX has an extra average jitter over Ryu controller. Figure 15 shows that when POX and Ryu controllers have 10 Mb/s and 1Mb/s of link bandwidth respectively, POX has a longer variation in average jitter than Ryu. Also it is obvious that the values of the average delay and jitter increase as the payload increases for both controllers and for all bandwidth cases.

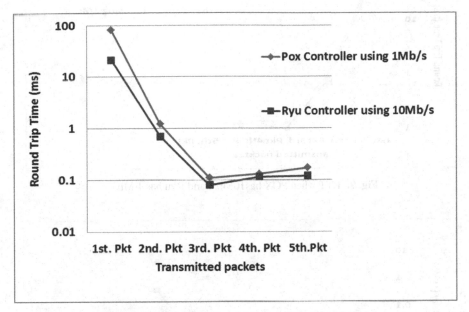

**Fig. 7.** RTT when POX has 1 Mb/s and Ryu has 10 Mb/s.

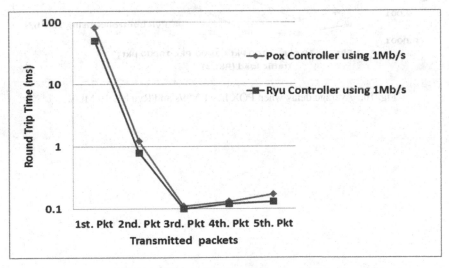

**Fig. 8.** RTT when POX and Ryu has 1 Mb/s.

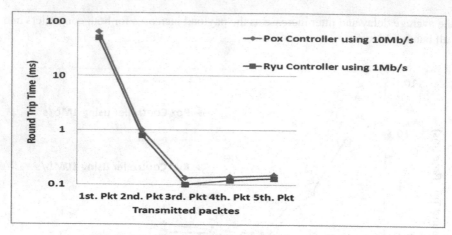

**Fig. 9.** RTT when POX has10Mb/s and Ryu has 1 Mb/s.

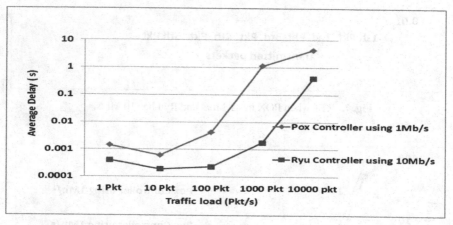

**Fig. 10.** Average delay when POX has 1 Mb/s and Ryu has 10 Mb/s.

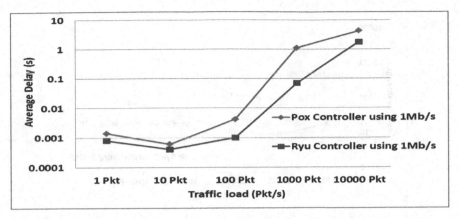

**Fig. 11.** Average delay when POX and Ryu have 1 Mb/s.

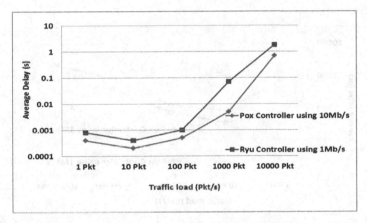

**Fig. 12.** Average delay when POX has10 Mb/s and Ryu has 1 Mb/s.

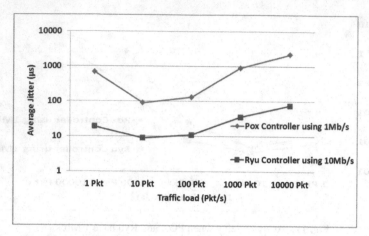

**Fig. 13.** Average jitter when POX has 1 Mb/s and Ryu has10 Mb/s.

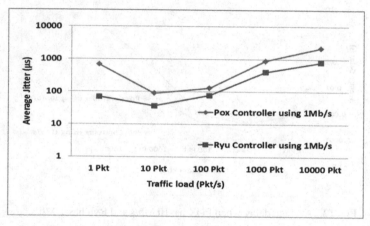

**Fig. 14.** Average jitter when POX and Ryu have 1 Mb/s.

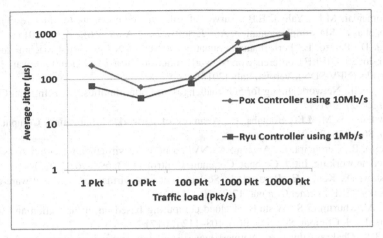

**Fig. 15.** Average jitter when POX has 10 Mb/s and Ryu has 1 Mb/s.

## 5 Conclusions

Virtual SDN is beneficial for many applications such as security, Cloud computing, data centres etc. Three important parameters were analyzed between the end hosts for each slice, which are average delay, average jitter and Round Trip Time.

According to the emulation results, it is concluded that comparing the average delay and the average jitter for three types of network topology shows that the linear topology has the least values, then the star topology and lastly the diamond topology for the two controllers. As compared to POX, Ryu controller reduced the RTT of the first transmitted packet by 68%, 73%, and 62% for the linear, star and diamond topology respectively.

Also, the results confirm that as the number of shared switches between the two slices increase, the delay factor becomes more effective, as in linear topology there was no shared switches, while in star topology and diamond topology the shared switches were one and two respectively.

Furthermore; the results show that Ryu controller achieved better performance as compared to POX controller especially for large amounts of payload, due to the properties and functions of Ryu controller used to perform network topology discovery. Ryu is able to reduce the round trip time by 31% at 10 k (pkt/Sec) as compared to POX when both controllers have the same link bandwidth. It can be seen that the RTT using Ryu controller is 74% less than POX controller when Ryu has 10 Mb/s and POX has 1 Mb/s of link bandwidth, and that Ryu reduced the average delay by 58% at 10 k (pkt/Sec).

Also, we noticed that both the average delay and average jitter exponentially increase as the payload increases for all test cases.

## References

1. Jain, A., Gupta, A., Gedia, D., Perigo, L., Gandotra, R., Murthy, S.: Trend-based networking driven by big data telemetry for SDN and traditional networks. Int. J. Next-Gener. Netw. (IJNGN) **11**(1) (2019)

2. Al-Somaidai, M.B., Yahya, E.B.: Survey of software components to emulate OpenFlow protocol as an SDN implementation. Am. J. Softw. Eng. Appl. **3**(6), 74–82 (2014)
3. Gedia, D., Perigo, L.: Performance evaluation of SDN-VNF in virtual machine and container. In: 2018 IEEE Conference on Network Function Virtualization and Software Defined Networks (NFV-SDN), Verona, Italy (2018)
4. Li, X., et al.: Network slicing for 5G: challenges and opportunities. IEEE Internet Comput. **21**(5), 20–27 (2017)
5. Chowdhury, N.M.M.K., Boutaba, R.: A survey of network virtualization. Comput. Netw. **54**(5), 862–876 (2010)
6. Pinheiro, B., Cerqueira, E., Abelem, A.: NVP: a network virtualization proxy for software defined networking. Int. J. Comput. Commun. Control **11**(5), 697–708 (2016)
7. Drutskoy, D., Keller, E., Rexford, J.: Scalable network virtualization in software-defined networks. IEEE Internet Comput. **17**(2), 20–27 (2013)
8. Kaur, A., Khurmi, S.S.: A study of cloud computing based on virtualization and security threats. Int. J. Comput. Sci. Eng. **5**(9), 108–112 (2017)
9. Raol, P.S., Chakravarthy, C.K.: Network hypervisors pros and cons: a survey. Int. J. Comput. Sci. Eng. **5**(11) (2017)
10. Yu, M., Rexford, J., Sun, X., Rao, S., Feamster, N.: A survey of virtual LAN usage in campus networks. IEEE Commun. Mag. **49**(7), 98–103 (2011)
11. Salvadori, E., Corin, R.D., Broglio, A., Gerola, M.: Generalizing virtual network topologies in OpenFlow-based networks. In: Proceeding of IEEE Globecom, Houston, Texas, USA, pp. 1–6 (2011)
12. Corin, R.D., Gerola, M., Riggio, R., De Pellegrini, F., Salvadori, E.: VeRTIGO: network virtualization and beyond. In: Proceeding of European Workshop on Software Defined Networks (EWSDN), Trento, Italy, pp. 24–29 (2012)
13. Jin, X., Gossels, J., Rexford, J., Walker, D.: CoVisor: a compositional hypervisor for software-defined networks. In: Proceeding in USENIX Symposium on Networked Systems Design and Implementation (NSDI), Barcelona, Spain, pp. 87–101 (2015)
14. Sherwood, R., et al.: FlowVisor: a network virtualization layer. Deutsche Telekom Inc. R&D Lab, Stanford, Nicira Networks, Technical report (2009)
15. Sherwood, R., et al.: Can the production network be the testbed? In: Proceeding of the Operating Systems Design and Implementation (OSDI), Vancouver, BC, Canada, October (2010)
16. Rao, P.S., Chakravarthy, C.K.: Network hypervisors pros and cons: a survey. Int. J. Comput. Sci. Eng. **5**(11) (2017)
17. Blenk, A., Basta, A., Reisslein, M., Kellerer, W.: Survey on network virtualization hypervisors for software defined networking. IEEE Commun. Surv. Tut. **17**(1), 27–51 (2015)
18. Pfaff, B., Pettit, J., Koponen, T., Amidon, K., Casado, M., Shenker, S.: Extending networking into the virtualization layer. In: Proceeding of the 8th ACM Workshop on Hot Topics in Networks (HotNets-VIII), UC Berkeley (2009)
19. Gutz, S., Story, A., Schlesinger, C., Foster, N.: Splendid isolation: a slice abstraction for software-defined networks. In: Proceeding of the 1st Workshop on Hot Topics in Software Defined Networks, Helsinki, Finland (2012)
20. Ryu controller. http://osrg.github.io/ryu/
21. POX controller. http://www.noxrepo.org/pox/about-pox/
22. Flowvisor controller. https://github.com/OPENNETWORKINGLAB/flowvisor/wiki

# Brain Computer Interface Enhancement Based on Stones Blind Source Separation and Naive Bayes Classifier

Zainab Kadham Abass[1]([⊠]), Taha Mohammed Hasan[1]([⊠]),
and Ahmed Kareem Abdullah[2]([⊠])

[1] College of Sciences, University of Diyala, Baqubah, Diyala, Iraq
zainabkadam10@gmail.com, Dr.tahamh@sciences.uodiyala.edu.iq
[2] Technical College-AL Mussaib, Al-Furat Al-Awsat Technical University, Babil, Iraq
Com.ahmd7@atu.edu.iq

**Abstract.** In neuroscience, Electroencephalography (EEG) can be considered as an extensively recognized process, implemented for the purpose of extracting brain signal activities related to involuntary and voluntary tasks. The researchers and scientists in the neuroscience field are concerned with studying the brain-computer interfacing (BCI) as well as enhancing the current systems of BCI. In the brain-computer interfacing area, Brain-Computer Interface (BCI) can be defined as a communication system that is developed for allowing individuals experiencing complete paralysis sending commands or messages without sending them via the normal output path-way of the brain. This study implements algorithms which can separate and classify task-related Electroencephalography (EEG) signals from ongoing EEG signals. The separation was made by hybridization between the classical method represented by the filtering process and modern method represented by Stone's BSS technique assessed to their capability of isolating and deleting electromyography (EMG), electrooculographic (EOG), (ECG) and power line (LN) artifacts into individual components that are considered artifacts to affect the performance of the system, and task-related Electroencephalography (EEG) signal are classified by Naïve Bayes (NB). The obtained results of the recognition rate were 82% by proposed algorithms. The results are compatible with previous studies.

**Keywords:** Brain Computer Interface (BCI) · Blind source separation · Naive Bayes (NB)

## 1 Introduction

The BCI is a direct connection between a human or animal brain (or culture of the brain cells) and external device, which can directly conduct some external activities according to the signals from the brain. Prior to this, a research team at Tsinghua University developed a series of internationally advanced methods for signal processing and pattern classification, they used this series of methods to successfully implement the BCI for

© Springer Nature Switzerland AG 2020
A. M. Al-Bakry et al. (Eds.): NTICT 2020, CCIS 1183, pp. 17–28, 2020.
https://doi.org/10.1007/978-3-030-55340-1_2

real-time brain activity cortical signal interpretation [1]. Additional development related to the systems of BCI is based on the improvement of novel training approaches for the patients, removing the noise that interfere with the signals and identifying the signals most appropriate for voluntary control. EEG recording artifacts could be a result of cardiac noise, blinks, eye movement, in addition to non-biological sources (power-line noise). There will be an obstacle if the subject generates an artifact which will be utilized for controlling the system of BCI since it will violate the specification of BCI as a non-muscular communication channel. Moreover, the ability of subjects suffering degenerative diseases could be lost. For example, such artifacts can be involuntary or voluntary blinks or contractions of the muscle when presenting the task. Furthermore, ocular activity or involuntary muscle could obscure actual EEG signal, blocking the measurement of features utilized for controlling the system [2].

Essentially, BSS algorithms examine some alternatives and in addition to some encouraging methods. The algorithms have been utilized not just for removing artifacts or reducing noise, but also in the source localization and enhancing the spatial resolution regarding the signals of the brain [3].

BCI could be referred to as Brain Machine Interface (BMI), Mind-Machine Interface (MMI), or direct neural interface. These terms lead to recognizing BCI as a direct communication tube between an external device and the brain as depicted in Fig. 1. Furthermore, the system can be defined as a major technology in the years to come that have the aim of using the person's thoughts and abilities to implement the intentions of individuals to the world [4].

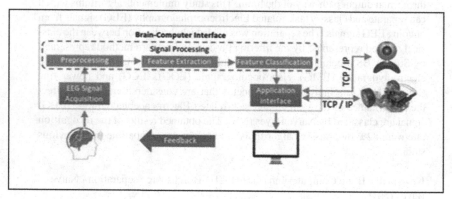

**Fig. 1.** The components regarding the BCI-based system [5].

Several researchers have shown their interest in brain-computer interface system since the importance of BCI system lies in many applications, such as medical applications especially. The following are some of the published works that are relevant to the current work.

**In** [6]: talk about BCI for Control of Wheelchair through the use of Fuzzy Neural Network. They used a Fast Fourier Transform (FFT) to extract important features from the EEG signal. Then the extracted features are input signals of the FNN based classifier, they found that the use of FFN is very effective in the classification of EEG signals.

**In** [7]: The researchers compared the different parameter settings systematically to extract wavelet-based features for optimum performance, which is an important step in Motor Imagery (MI) based BCI, which translates the brain signals into external control commands. They reached results through their detailed experiments of the possibility of selecting (the function of the appropriate wavelets, their order, the number of levels of decomposition and the selection of laboratories in the end at different levels).

**In** [8]: traditional electrodes have been used by the researchers, these electrodes were attached around the ear in a distance about 1.5 cm from the ear, the 2-class MI tasks have been classified by the researchers through the use of ear-around EEG, they also suggested common spatial pattern (CSP)-based frequency-band algorithm of optimization and put it to comparison with 3 current approaches.

**In** [9]: the researchers proposed analytic intrinsic mode functions (AIMFs) for the classification of EEG signals of various MI tasks. In which AIMFs will be acquired through using Hilbert transform and empirical mode decomposition (EMD) on EEG signals. Those features are: raw moment which is related to the first instantaneous frequency derivative, area, a spectral moment of power spectral density, and peak value of PSD is estimated from AIMF. Normalized features will be utilized as inputs to least squares SVM (LS-SVM) classifier. The performance parameters have been estimated through the use of different kernel functions of the classifier of LS-SVM.

The remnants of the study have been arranged in the following way: We touch the important topic in this research in the next section. In Sect. 3, the suggested model will be discussed. The experimental results will be presented in Sect. 4. Lastly, the conclusions of the study will be presented in Sect. 5.

## 2  Blind Source Separation (BSS)

This method is used for separating sources of signals without prior knowledge or may have very simple information about sources [10]. Figure 2 illustrates the diagram of BSS technique.

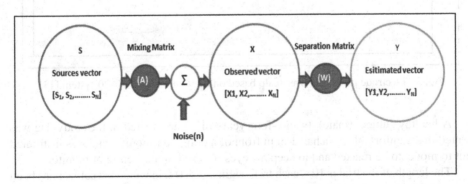

**Fig. 2.** Blind source separation stages [11].

The mixing equation is:

$$X(n) = AS(n) \tag{1}$$

The extracted equation is:

$$Y(n) = WX(n) \tag{2}$$

Where:

S(n) is Original source, X(n) is Observed signal, Y(n) is Estimated source, A is Mixing Matrix and W is Separation Matrix.

## 3 The Proposed Classification System

### 3.1 Dataset

The data have been acquired from [12]. EEG signals have been estimated through the use of computerized EEG device. The signals have been preserved as a data-base for additional processing. The scalp has been covered by nineteen electrodes based on 10–20 international system, as depicted in Fig. 3, recorded signals have been digitized at 256 Hz based on the specifications of computerized EEG device.

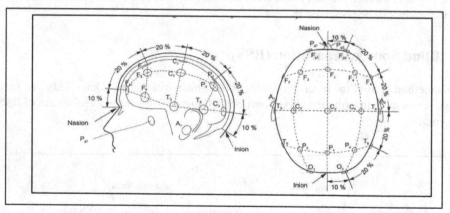

**Fig. 3.** Electrode placement over scalp based on the international 10–20 system [13].

A healthy subject (male), twenty-four years old, participated in the study. He was seated in a comfortable armchair 1 m in front of a computer monitor and was instructed not to move, to be relaxed and to keep his eyes focused on the centre of monitor.

The length of the trial is 10 s, each trial begins with 0 s, in which the subject has been permitted to do random artifacts like eye blinking. At second 3, the subject must stay calm and does not perform any activity. From second 5 to second 6, the subject achieved right or left hand index movement according to the part of session as displayed in Fig. 4.

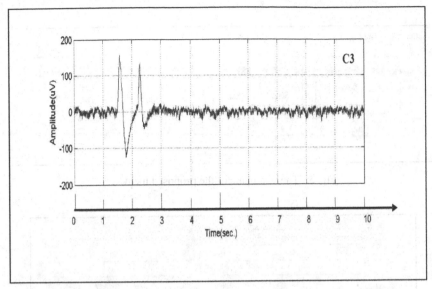

**Fig. 4.** One trial recorded at electrode C-3 when a movement of the right index finger was performed by the volunteer of 24 years old

## 3.2 Processing and Proposed Algorithms

The suggested Signal Processing in the presented study include 2 phases: the first phase is starting from preprocessing step represented by filtering process and modern method represented by Stone's BSS technique for feature extraction and the deletion of artifacts, since the important frequency limit is known according to the application of the BCI system for the movement of right finger and left finger between (Alpha ($\alpha$). They range from 8 to 12 Hz and Beta ($\beta$)) Frequency from 12 to 30 Hz, to focus only on a specific parameter and ignore frequencies Other. The second phase: takes the output from the first phase as input to Naïve Bayes (NB) Classifier to classify the signal and prediction to move the finger. Figure 5 represents the proposed method, and BPF neglect many frequencies which will grant BSS the ability to focus on a particular part of the signal and classify it by Naïve Bayes (NB) Classifier. Figure 6 shows the input signals part of the signal.

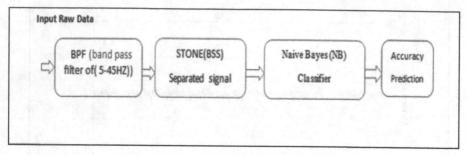

**Fig. 5.** The schematic of the proposed method.

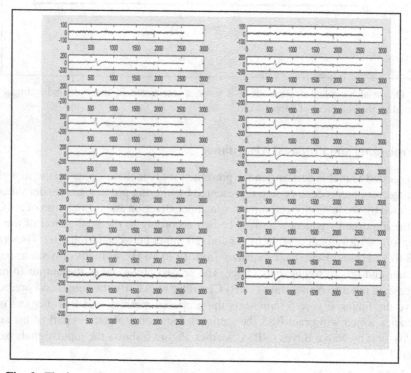

**Fig. 6.** The input signal part for 19 channel one trial of lift index finger movement

### 3.2.1 Preprocessing

A major issue related to automated analysis of EEG is detecting with the use of various types of interference wave-forms (i.e. artifacts) which have been added to the signal of EEG throughout the sessions of recording [14]. Detecting and eliminating artifacts in EEG is very complex, yet it is of high importance to develop practical systems. Artifacts can be defined as undesirable signals in EEG. They have different origins, that consist of utility frequency (60 Hz in USA or 50 Hz in Europe) [15]. Eye and body movement, or eye blinks, the utility frequency artifacts have been eliminated from data through the use of band pass filter.

### 1-Band Pass Filter (BPF)

Filter's name can give an idea about its work. BPF operate to pass band of frequencies from the signal and reject the others. Like any filter BPF needs to set first and second cut off frequency (fc1, fc2), and that needs to know the minimum and maximum allowed frequencies. The initial phase to process EEG signal is band-pass filtered at 5 Hz–45 Hz for removing DC. Drift, eyes artifacts, and power line noise of 50 Hz. A 5 Hz–45 Hz band-pass has been carried out through the use of Windowed-Since FIR filter with sampling rate (256 samples/s), in addition to filter kernel length M of 1024 computed based on the Eq. (3). The filter kernel regarding low-pass filter has been estimated depending on Eq. (4) [12].

$$M \approx \frac{4}{BW} \tag{3}$$

$$h[i] = K\frac{\sin(2\pi f_c(i - M/2))}{i - M/2}\left[0.42 - 0.5\cos\left(\frac{2\pi i}{M}\right) + 0.08\cos\frac{4\pi i}{M}\right] \tag{4}$$

where h[i] can be considered as the filter kernel.

K will be defined as the filter gain.

M will be defined as the length of filter kernel.

fc will be defined as the cutoff frequency as a ratio to the rate of sampling, and it will be defined as the index.

The next step would be entering the mixed signals into BPF to cut the desired part of frequencies from the mixed signals. Figure 7 below illustrates the response of BPF.

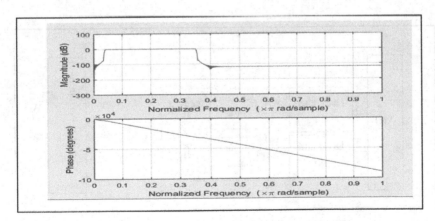

**Fig. 7.** BPF response

Figure 8 below shows the shape of mixed signal after passing through BPF. As mentioned previously the difference between signals out of mixing matrix and the output of BPF is very clear. The out of BPF will be the input to STONE BSS algorithm.

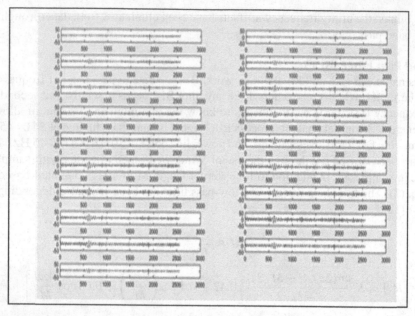

**Fig. 8.** The out signal for 19 channel after BPF one trial of lift index finger movement

## 2-Stone BSS Algorithm

This method of BSS utilizes the temporal predictability to separate sources from the mixture. The method also took the same name of the scientist who invented it [16]. The out of Stone BSS algorithm will be the input to Naive Bayes Classifier. Figure 9 represents Stone BSS.

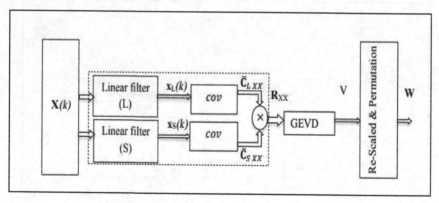

**Fig. 9.** Schematic diagram of the Stone BSS [16].

Where:

X (k) represents the Mixture observation signals

XS (k) represents the Filter Response (S)

XL (k) represents the Filter Response (L)
$\bar{C}LXX$ represents the Long-term covariance matrix
$\bar{C}SXX$ represents the Short-term covariance matrix
RXX represents the $\bar{C}LXX\tilde{C}SXX$
V represents the Eigenvector matrix RXXV = VD;
W represents the matrix of Un-mixing.
Stone's measure is indicated by defined as [11]:

$$F(y) = log\frac{Vy}{Uy} = log\frac{\sum_{k=1}^{N}(yl(k) - y(k))^2}{\sum_{k=1}^{N}(ys(k) - y(k))^2} \tag{5}$$

$$y_s(k) = \beta_s y_s(k-1) + (1 - \beta_s)y(k-1) \tag{6}$$

$$y_l(k) = \beta_l y_l(k-1) + (1 - \beta_l)y(k-1) \tag{7}$$

According to Stone's method, βs. βL ∈ [0, 1] are two distinct parameters and yl(1) = ys(1) = y(1). Half-life hL of βL is much longer (usually × 100 longer) compared to the corresponding half-life hS of βs and the formula is:

$$\beta = 2^{\frac{-2}{h}} \tag{8}$$

Where N is the max No. samples signal y(k).

### 3.2.2 Classification

In the presented study, the data will be classified through the use of Naïve Bayes Classifier.

**Naive Bayes Classifier**
Naive Bayes method is used in the presented paper, because of its ease and transparency in the modalities of ML. This method is basically done according to Bayesian theorem with assumption of strong independence amongst features [17].

$$P(C/X) = \frac{P(x|c)P(c)}{P(x)} \tag{9}$$

Where P(c/x) can be defined as the feature probability of class (i.e. target) of a certain feature, P(c) can be defined as the prior probability of class, P(x | c) can be defined as the likelihood that is the likelihood of feature given class, and P(x) is the feature's prior likelihood.

## 4 Experimental Results

This section discussed the results when we used STONE BSS algorithm and a Naive Bayes Classifier in the brain interface system between the computer, and comparing these results when raw data is classified without filtering. Stone algorithm restored source signals successfully as shown in Fig. 10.

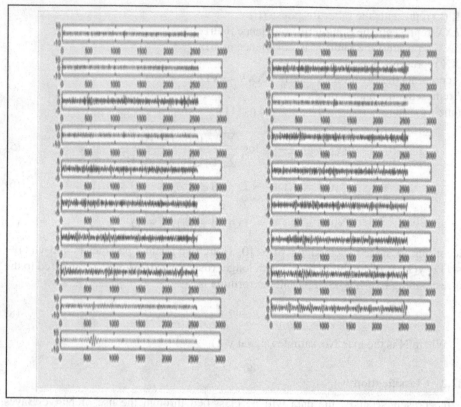

**Fig. 10.** Recover Sources By Stone one trial of lift index finger movement (proposed work)

**Table 1.** Results of classification accuracy using proposed techniques

|  | Preprocessing | Classification | Precision | Recall | F-Measure |
|---|---|---|---|---|---|
| EEG data | Without processing | Naïve Bayes | 56% | 51% | 40% |
|  | With STONE |  | 79% | 74% | 72% |
|  | BPF with STONE proposed work |  | 82% | 82% | 81% |

Table 1 represents the results of the proposed work and Table numbers showed that the proposed work has higher performance and higher rating accuracy.

Generally, proposed work has better performance than other algorithms by experiment whit for this reason STONE has been selected to be improved since it is already more efficient than the other used blind source separation algorithms. The results gave a higher rating accuracy, as shown in Fig. 11.

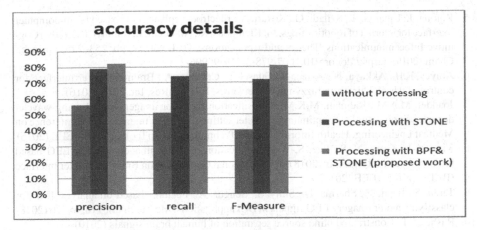

**Fig. 11.** Accuracy using proposed techniques.

## 5 Conclusions

In the presented study, a new approach is suggested based on connected the BPF with Stone blind source separation technique, an algorithm that was first used in the brain-computer interface system in our research to extract and separate the useful information and has been used for reducing the range of frequency and delete artifacts that affect the performance of the system rating, also used a classification model Naive Bayes algorithm. The obtained results of the recognition rate were 82% by Proposed algorithms. The results are compatible with previous studies. We concluded that the proposed system using a blind source algorithm with Naive Bayes algorithm leads to high efficiency and performance because high noise affects system performance and the hybridization process between band pass filter and the stone algorithm led to high precision rate.

The possible future works for BCI take several directions and thus, suggest developing an algorithm to determine the reactive frequency of beta band and to select it automatically and extended the proposed algorithm for multiclass MI classification purpose and increasing the number of mental activities such as foot and tongue movements.

## References

1. Andruseac, G.G., Paturca, S.V., Banica, C.K., Costea, I.M., Rotariu, C.: A novel method of teaching information technology applied in health monitoring. J. Biotechnol. **239**, 1–3 (2016)
2. McFarland, D.J., Sarnacki, W.A., Vaughan, T.M., Wolpaw, J.R.: Brain-computer interface (BCI) operation: signal and noise during early training sessions. Clin. Neurophysiol. **116**, 56–62 (2005)
3. Cichocki, A.: Blind signal processing methods for analyzing multichannel brain signals. System **1**, 2 (2004)
4. Mousa, F.A., El-Khoribi, R.A., Shoman, M.E.: An integrated classification method for brain computer interface system. In: 2015 Fifth International Conference on Digital Information Processing and Communications (ICDIPC), pp. 141–146. IEEE (2015)

5. Katona, J., Ujbanyi, T., Sziladi, G., Kovari, A.: Electroencephalogram-based brain-computer interface for internet of robotic things. In: Klempous, R., Nikodem, J., Baranyi, P.Z. (eds.) Cognitive Infocommunications, Theory and Applications. TIEI, vol. 13, pp. 253–275. Springer, Cham (2019). https://doi.org/10.1007/978-3-319-95996-2_12

6. Abiyev, R.H., Akkaya, N., Aytac, E., Günsel, I., Çağman, A.: Brain-computer interface for control of wheelchair using fuzzy neural networks. Biomed Res. Int. **2016** (2016)

7. Joadder, M.A.M., Rahman, M.K.M.: Classification of motor imagery signal using wavelet decomposition: a study for optimum parameter settings. In: 2016 International Conference on Medical Engineering, Health Informatics and Technology (MediTec), pp. 1–6. IEEE (2016)

8. Kim, Y.-J., Kwak, N.-S., Lee, S.-W.: Classification of motor imagery for Ear-EEG based brain-computer interface. In: 2018 6th International Conference on Brain-Computer Interface (BCI), pp. 1–2. IEEE (2018)

9. Taran, S., Bajaj, V., Sharma, D., Siuly, S., Sengur, A.: Features based on analytic IMF for classifying motor imagery EEG signals in BCI applications. Measurement **116**, 68–76 (2018)

10. Rasheed, T.: Constrained blind source separation of human brain signals (2010)

11. Abdullah, A.K., Wadday, A.G., Abdullah, A.A.: Separation enhancement of power line noise from human ECG signal based on stone technique. In: Journal of Biomimetics, Biomaterials and Biomedical Engineering, pp. 71–78. Trans Tech Publ. (2019)

12. Hussein, S.M.: Design and implementation of AI controller based on brain computer interface, Master. Computer Engineering, Nahrain University, Iraq-Baghdad (2007)

13. GomeNicolas-Alonso, L.F., Z-Gil, J.: Brain computer interfaces, a review. Sensors **12**, 1211–1279 (2012)

14. Sleight, J., Pillai, P., Mohan, S.: Classification of executed and imagined motor movement EEG signals. Ann Arbor, Univ. Michigan, pp. 1–10 (2009)

15. Fatourechi, M., Bashashati, A., Ward, R.K., Birch, G.E.: EMG and EOG artifacts in brain computer interface systems: a survey. Clin. Neurophysiol. **118**, 480–494 (2007)

16. Abdullah, A.K., Zhu, Z.C.: Enhancement of source separation based on efficient stone's BSS algorithm. Int. J. Signal Process. Image Process. Pattern Recognit. **7**, 431–442 (2014)

17. Naseer, N., Qureshi, N.K., Noori, F.M., Hong, K.-S.: Analysis of different classification techniques for two-class functional near-infrared spectroscopy-based brain-computer interface. Comput. Intell. Neurosci. **2016** (2016)

# Policing Based Traffic Engineering Fast ReRoute in SD-WAN Architectures: Approach Development and Investigation

Oleksandr Lemeshko[1] , Oleksandra Yeremenko[1]([✉]) , Ahmad M. Hailan[2] ,
Maryna Yevdokymenko[1] , and Anastasiia Shapovalova[1]

[1] Kharkiv National University of Radio Electronics, 14 Nauky Ave., Kharkiv, Ukraine
{oleksandr.lemeshko.ua,oleksandra.yeremenko.ua,
maryna.yevdokymenko}@ieee.org, anastasiia.shapovalova@nure.ua
[2] Thi-Qar University, Nasiriya, Iraq
ahmad.m.hailan@utq.edu.iq

**Abstract.** The paper is devoted to the approach development and investigation of the policing based Traffic Engineering Fast ReRoute in SD-WAN architectures. The paper proposes a mathematical model of Fast ReRoute with load balancing based on the principles of Traffic Engineering (TE) and differentiated traffic policing in communication networks. The basis of the model is the conditions for the implementation of multipath routing together with the modified conditions for flow conservation, which take into account the priority traffic policing on the network edge in case of its probable overload. On the one hand, such an overload is caused by the increase in the load. On the other hand, it is caused by the implementation of protection schemes for network elements and its bandwidth in the course of fast rerouting. The advantage of the proposed solution is also the formulation of the Traffic Engineering Fast ReRoute with support of Traffic Policing (TE FRR-TP) as an optimization one. The given optimality criterion focuses firstly on minimizing the dynamically controlled upper bound of communication links utilization that meets the requirements of the TE concept. Secondly, it minimizes the intensity of flows that receive denial of service at the edge of the network weighted in relation to the priority of serving. Using the proposed model on a number of numerical examples, the research of the policing based Traffic Engineering Fast ReRoute processes has confirmed the adequacy and efficiency of routing solutions made on its basis, both in terms of ensuring their fault-tolerance and load balancing, and in relation to traffic policing based on priorities.

**Keywords:** SD-WAN · Networks resilience · Traffic Engineering · Fast ReRoute · Traffic policing · Bandwidth protection

## 1 Introduction

Currently, the use of such a concept as software-defined networking in a wide area network (SD-WAN) is a promising technological solution that allows consolidating network functions and services along with simplification of infrastructure and management

© Springer Nature Switzerland AG 2020
A. M. Al-Bakry et al. (Eds.): NTICT 2020, CCIS 1183, pp. 29–43, 2020.
https://doi.org/10.1007/978-3-030-55340-1_3

[1–3]. In this case, network configuration is separated from the underlying networking services provided by network service providers, thereby achieving effective bandwidth utilization, and saving time for deployment and reconfiguration of the wide-area network (WAN) according to the demands of user flows. Various technological solutions, policies, and smart routing algorithms that can be used in SD-WAN in order to help to adapt the network for differentiated servicing of user flows [1–8].

In the general case, according to the requirements of modern user applications, the following benefits of SD-WAN can be formulated:

- Optimal utilization of network resource (bandwidth), in which SD-WAN will choose the most suitable transport technology among the available ones for a certain application that generates a flow of data to be transmitted in the network;
- Manageability that reduces the cost of administering and managing the network unlike traditional networks;
- Network security, in which the encryption of data flows during their transmission provides protection when using any transport technology;
- Flexibility and scalability when reallocating SD-WAN bandwidth in differentiated service flows of users generated by relevant applications, including critical ones, which should receive guaranteed network service;
- High performance of SD-WAN, which can be equated to MPLS networks.

It should be noted that when choosing an SD-WAN-based technological solution, it is necessary to ensure a high level of its fault-tolerance to possible failures in network hardware or software, overload or violation of the information security level, etc. An important technological tool for increasing the reliability of network solutions is fault-tolerant routing, which at the level of the transport network (core) is implemented using the protocols of Fast ReRoute (FRR) [9–15].

During fast rerouting, increasing the fault-tolerance of a network with the support of the basic schemes for protecting its elements (link/node/path/bandwidth) is provided by the introduction of resource redundancy. However, this in turn negatively affects network performance and Quality of Service (QoS) level in general. Therefore, on the one hand, it is important to efficiently balance the use of the available network resource on the principles of Traffic Engineering (TE) [13–22]. On the other hand, in the case of probable network overload, it is necessary to apply a function of tools such as Traffic Policing (TP) [11, 12]. It should be noted that providing a consistent solution to load balancing and fault-tolerant routing tasks, as a rule, leads to increased computational complexity and reduced scalability of protocol solutions. It is known that the efficiency of the protocol solution is largely determined by the adequacy and quality of the mathematical model of the calculation laid down in its basis. As shown by the analysis [23], the order of FRR and TE is determined in the course of solving optimization problems of different levels of complexity.

## 2 Overview of Solutions on Fast ReRoute with Load Balancing in Software-Defined Wide Area Networks

Among research on fault-tolerance in SDN networks, let us emphasize the works [24–27]. For example, in [24], an algorithm for Local Fast Reroute (LFR) with flow aggregation in software-defined networks is proposed. In the LFR algorithm, in case of failure detection in a communication link, all traffic flows affected by a denial are aggregated into the so-called "large" flow. Next, the local backup path for rerouting is dynamically deployed by the SDN controller for the aggregated flow. Thus, the LFR algorithm reduces the number of current operations between the SDN controller and switching equipment. The numerical results have proven that the LFR provides fast recovery, minimizing the total number of flows in SDN.

The growing complexity of modern network applications and the huge demand for Internet resources require SD infrastructures to adapt to the demands of a high degree of robustness and reliability. As mentioned above, in SDN networks the crucial task is increasing of fault-tolerance and timely updating of network state information, to which research [25] is dedicated. It identifies new algorithms designed to improve the search for backup paths in large-scale networks with single failures of communication links with minimal time costs for updating network state information. The new solution aims to increase the efficiency and reduce the processing of service information in case of denial of communication links.

The Adaptive Dynamic Multi-Path Computation Framework (ADMPCF) was proposed in [27] to provide efficient management of network resources for organizing routing and resource allocation under the condition of centralized management of a wide software-defined network. Such a system can provide the necessary infrastructure for integrating data and analytics gathering, evaluating network performance using different optimization algorithms.

It should be noted that, as a rule, implementation of the scheme of network bandwidth protection leads to a nonlinear formulation of the optimization problem and the corresponding increase in the computational complexity of the resulting solutions. In turn, the traffic management flow-based model was proposed in [28, 29], where a solution of the multipath routing with the traffic policing together with dynamic bandwidth allocation for the user flows was provided.

Thus, the scientific and practical task, which is related to ensuring the agreed solution of such rather complicated network tasks as FRR, TE, and TP, is based on the development of an adequate mathematics computational model. Based on the experience of developing and researching QoS routing models with TP support [28, 29], models and methods of FRR and TE FRR [30–35], the key requirements, which the new mathematical model TE-FRR-TP must satisfy, can be formulated:

- accounting of specifics of the processes in networks;
- support of different routing schemes;
- implementation of known fault-tolerance protection schemes based on network elements redundancy as well as its bandwidth;
- acceptable computational complexity in subsequent protocol implementation.

## 3    Flow-Based Model of Traffic Engineering Fast ReRoute with Support of Policing

Suppose that the structure of communication network is defined as the graph $G = (R, E)$, where $R = \{R_i; \; i = \overline{1, m}\}$ is the set of routers in the communication network, and $E = \{E_{i,j}; \; i, j = \overline{1, m}; \; i \neq j\}$ is the set of links. Let us denote by $R_i^* = \{R_j : E_{j,i} \neq 0; \; j = \overline{1, m}; \; i \neq j\}$ the subset of routers incident to the $R_i$ router. Then the number of communication links in the network is defined by $n = |E|$, and the capacity $\varphi_{i,j}$ will be assigned to each of the links.

Within the model, a number of functional parameters are associated with each $k$th flow: $s_k$ is the source router; $d_k$ is the destination router; $\lambda^k$ is the average intensity (packet rate) of the $k$th flow in packets per second (1/s); $K$ is the set of flows transmitted in the network ($k \in K$).

The result of solving the problem of Traffic Engineering Fast ReRoute with support of policing is computation of two types of routing variables $x_{i,j}^k$ and $\bar{x}_{i,j}^k$ that characterize the proportion of the intensity of the kth flow within the link $E_{i,j} \in E$, forming the primary or backup path.

In the case when the multipath strategy for routing variables of two types is used in the communication network, the following constraints take place [30, 31]

$$0 \leq x_{i,j}^k \leq 1 \text{ and } 0 \leq \bar{x}_{i,j}^k \leq 1. \tag{1}$$

The flow conservation conditions ensuring the connectivity of the calculated routes are introduced for the routing variables of the primary path [28]

$$\begin{cases} \sum_{j:E_{i,j}\in E} x_{i,j}^k - \sum_{j:E_{j,i}\in E} x_{j,i}^k = 0; \, k \in K, R_i \neq s_k, d_k; \\ \sum_{j:E_{i,j}\in E} x_{i,j}^k - \sum_{j:E_{j,i}\in E} x_{j,i}^k = 1 - \beta^k; \, k \in K, R_i = s_k; \\ \sum_{j:E_{i,j}\in E} x_{i,j}^k - \sum_{j:E_{j,i}\in E} x_{j,i}^k = \beta^k - 1; \, k \in K, R_i = d_k; \end{cases} \tag{2}$$

where $\beta^k$ is the proportion of the intensity of the $k$th flow that receives a denial of service when using the primary path. The same conditions have to be imposed for the backup path routing variables [28]:

$$\begin{cases} \sum_{j:E_{i,j}\in E} \bar{x}_{i,j}^k - \sum_{j:E_{j,i}\in E} \bar{x}_{j,i}^k = 0; \, k \in K, R_i \neq s_k, d_k; \\ \sum_{j:E_{i,j}\in E} \bar{x}_{i,j}^k - \sum_{j:E_{j,i}\in E} \bar{x}_{j,i}^k = 1 - \bar{\beta}^k; \, k \in K, R_i = s_k; \\ \sum_{j:E_{i,j}\in E} \bar{x}_{i,j}^k - \sum_{j:E_{j,i}\in E} \bar{x}_{j,i}^k = \bar{\beta}^k - 1; \, k \in K, R_i = d_k; \end{cases} \tag{3}$$

where $\bar{\beta}^k$ is the proportion of the intensity of the $k$th flow that receives a denial of service when using the backup path.

## 4 Conditions of Link, Node, and Bandwidth Protection in Traffic Engineering Fast ReRoute with Support of Policing

In the course of Fast ReRouting with a realization of the link $E_{i,j} \in E$ protection scheme and using the multipath routing strategy over the backup paths (multipaths), the following condition is introduced [32]

$$0 \le \bar{x}_{i,j}^k \le \delta_{i,j}^k \qquad (4)$$

where

$$\delta_{i,j}^k = \begin{cases} 0, & \text{when protecting the link } E_{i,j}; \\ 1, & \text{otherwise.} \end{cases} \qquad (5)$$

Ensuring the linear conditions (4) and (5) guarantees that the link $E_{i,j} \in E$, which is being protected, will not be used in the backup path in the case of using the multipath routing.

In turn, when the node $R_i \in R$ is being protected, these conditions (4) and (5) can be used for the case of protecting a set of links adjacent to this node [32]. Here in multipath routing, the constraints have to be satisfied:

$$0 \le \bar{x}_{i,j}^k \le \delta_{i,j}^k \text{ under } R_j \in R_i^*, j = \overline{1,m} \qquad (6)$$

where the values of $\delta_{i,j}^k$ are determined according to (5); $R_i^*$ is a set of nodes that are adjacent to the node $R_i$, i.e. when $E_{i,j} \ne 0$.

It worth noting that meeting the mentioned above requirements (5) and (6) guarantees that the node $R_i \in R$ will be protected and the backup path will not use all the links outgoing from this node. Moreover, conditions (4) and not using the outgoing links prevent comprising of links incoming to protected transit nodes into the backup path. This is how the protection of the particular node $R_i$ is implemented as a whole.

The conditions for preventing overload in the network links during the Fast ReRoute implementation, when some flows can switch to the backup routes, have the general form:

$$\sum_{k \in K} \lambda^k \cdot \max\left[x_{i,j}^k, \bar{x}_{i,j}^k\right] \le \varphi_{i,j}, E_{i,j} \in E. \qquad (7)$$

The model proposes to introduce the following modified conditions for preventing overloading in order to provide load balancing in the network [35]:

$$\sum_{k \in K} \lambda^k \cdot u_{i,j}^k \le \alpha \cdot \varphi_{i,j}, E_{i,j} \in E \qquad (8)$$

under

$$x_{i,j}^k \le u_{i,j}^k \text{ and } \bar{x}_{i,j}^k \le u_{i,j}^k \qquad (9)$$

where $u_{i,j}^k$ are also control variables

$$0 \le u_{i,j}^k \le 1, \qquad (10)$$

which represent the upper bound values of routing variables of the primary and backup paths, and $\alpha$ is the additional control variable that numerically determines the upper bound of the network links utilization and follows the given conditions

$$0 \le \alpha \le \alpha_{TH} \tag{11}$$

where $\alpha_{TH}$ is the threshold of the upper bound of the network links utilization, the value of which is determined based on the requirements for the Quality of Service level in the network. This is due to the fact that all the main QoS indicators – network performance, average end-to-end delay and the packet loss probability are the function of the parameter $\alpha$. The introduction of conditions (11) is a novelty of the proposed approach compared to the models presented in [34, 35]. At the same time, maintaining the linearity of the model (1)–(11) and focusing on ensuring a given QoS level is one of the main advantages of the proposed solution.

An optimality criterion of the problem solutions of the Traffic Engineering Fast ReRoute with support of policing will be the minimum of the following function

$$J = \sum_{k \in K} w_k \cdot \beta^k + \sum_{k \in K} \bar{w}_k \cdot \bar{\beta}^k + c \cdot \alpha \to \mathbf{min} \tag{12}$$

where weighting coefficients must follow the given condition

$$w_k > \bar{w}_k > w_p > \bar{w}_p > \ldots > c, \tag{13}$$

provided that the priority of the $k$th flow $(PR^k)$ is higher than the priority of the $p$ th flow $(PR^p)$. In IP network when using the 3 bits of IP precedence in IP packet header for prioritization, the value of priority is of the range from 0 to 7. However, for the DSCP (Differentiated Services Code Point), the priorities will vary from 0 to 63. Then, according to (13), it is proposed to use the following weighting coefficients in the criterion (12):

$$w_k = PR^k + 1, \ \bar{w}_k = PR^k + 0.5, \ c = 0.25.$$

Then the optimality criterion (12) focuses on minimizing the conditional costs associated with the consistent solution of the FRR, TE, and TP tasks. In this case, the first term determines the conditional cost of denials to maintain flows being transmitted through the primary paths. The second term defines the conditional cost of denials of servicing the flows being transmitted in the backup paths. The third term describes a weighted upper bound of the network links utilization. The formulated optimization problem belongs to the linear programming class, which in the course of ongoing research was solved in the MATLAB environment using the `linprog` function from the `Optimization Toolbox`.

## 5  Numerical Example

The analysis of the proposed Traffic Engineering Fast ReRoute with support of the policing model was conducted on different network configurations for a varying number of flows and their characteristics. The features of the model are illustrated by the following numerical example. The structure of the investigated network is shown in Fig. 1. The capacities of network links are indicated in the gaps.

Suppose that there is a need to provide a solution to the problem of Traffic Engineering Fast ReRoute with support of policing with the implementation of the protection scheme of the $E_{11,12}$ link for two flows. It is assumed that the characteristics of the flows are the following:

- the first flow is transmitted from the node $R_1$ to the node $R_{16}$ with varying intensity $\lambda^1 = 10 \div 1100$ 1/s and $PR^1 = 4$;
- the second flow is transmitted from the node $R_5$ to the node $R_{12}$ with varying intensity $\lambda^2 = 10 \div 1100$ 1/s and $PR^2 = 1$.

The threshold of the upper bound of the network links utilization is supposed to be equal, for example, $\alpha_{TH} = 0.75$.

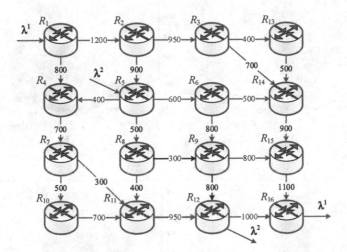

**Fig. 1.** Network structure for numerical example.

As shown by the research results presented in Fig. 2, with increasing of network load, the upper bound of the network links utilization also gradually increased. The absence of sharp fluctuations in values of $\alpha$ positively influences the quality of service in the network as a whole. At the same time, at low network load when $\lambda^1 \leq 900$ 1/s and $\lambda^2 \leq 830$ 1/s, fulfillment of condition (11) under $0 \leq \alpha < \alpha_{TH}$ did not lead to a limiting of the intensity of flows at the network edge, i.e. $\beta^1 = \bar{\beta}^1 = \beta^2 = \bar{\beta}^2 = 0$ (Fig. 2). However, in the case of excessive load on the network, the condition (11) was satisfied in such a way when $\alpha = \alpha_{TH}$ (Fig. 2a) by limiting the intensities of flows that occurred both in the primary and in the backup paths. As can be seen from Fig. 2, traffic limitation was based on two main principles:

- limitations particularly concerned the flow that was a source of overload under condition (11);
- if the overload created several flows, then the limitations primarily concerned the flow with a lower priority in accordance with the conditions (13).

These principles are confirmed by the fact that at the specified initial data, the first (high-priority) flow of packets when implementing the primary route was not limited in its intensity, i.e. $\beta^1 = 0$. The second (low-priority) flow was limited by using the backup path earlier than others and with greater intensity (Fig. 2d). Somewhat later and with less intensity, the second flow was limited by the use of the primary path (Fig. 2c). The first flow, which had a high priority, was limited only in the case of overloading the links included to the backup path (Fig. 2b).

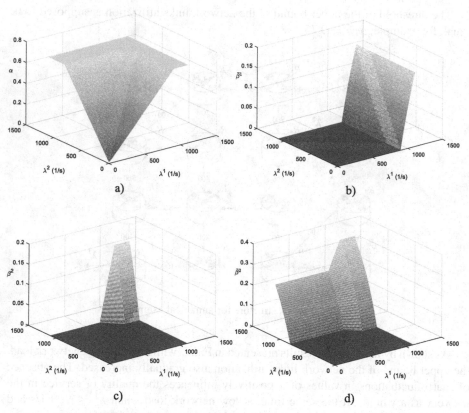

**Fig. 2.** The results of investigation for $\alpha_{TH} = 0.75$.

For clarity, let us look at the more detailed result obtained (Fig. 2) when $\lambda^1 = 950$ 1/s and $\lambda^2 = 1000$ 1/s. In Table 1, the results of solving the TE-FRR-TP problem for the two described flows are presented. The utilization $\alpha_{i,j}$ for each network link can be found as indicated in [35]:

$$\alpha_{i,j} = \frac{\sum_{k \in K} u_{i,j}^k \lambda^k}{\varphi_{i,j}} \tag{14}$$

Due to the network overload to ensure that condition (11) is met, it has been found that:

- when using the backup path, the first flow with high priority receives a denial of service at the network edge with the intensity 50 1/s, which corresponded to $\bar{\beta}^1 = 0.0526$;
- when using the primary path, the second flow with low priority receives a denial of service at the network edge with the intensity 100 1/s, which corresponded to $\beta^2 = 0.1$;
- when using the backup path, the second flow with low priority receives a denial of service at the network edge with the intensity 325 1/s, which corresponded to $\bar{\beta}^2 = 0.325$.

As shown by the numerical results, on the one hand, the model allows to provide denials (policing) based on differentiated services according to the priorities. On the other hand, it gives the possibility of limiting the flow that is the source of congestion. This was manifested in the fact that in case of overload, the shared links of the considered two flows ($E_{5,4}, E_{5,6}, E_{5,8}, E_{7,11}$, and $E_{8,9}$) were limited in intensity by the second (lower priority) flow (Table 1). Due to the probable nonfulfillment of condition (11) caused by the overload of the links unused by the second flow $E_{1,2}$ and $E_{14,15}$ (Table 1), the first flow (of high priority) was also limited, as shown in Fig. 2b.

Let us further consider the case when the threshold of the upper bound of the network links utilization is supposed to be lower and takes value equal to $\alpha_{TH} = 0.65$. Let us note that by reducing the threshold value $\alpha_{TH}$, the requirements for the Quality of Service increase. Numerical results are presented in Fig. 3 with the initial data corresponding to the previous example.

The obtained research results (Fig. 3a) confirmed the preliminary conclusion that the increase in the network load leads to a gradual increase in the upper bound of the network links utilization. As before, with a low network load ($0 \le \alpha < \alpha_{TH}$), there was no limitation of the intensity of flows at the network edge, that is $\beta^1 = \bar{\beta}^1 = \beta^2 = \bar{\beta}^2 = 0$. However, due to the fact that the threshold value $\alpha_{TH} = 0.65$ became more strict, traffic policing began at slightly lower values of the intensity of the input flows: $\lambda^1 > 780$ 1/s, $\lambda^2 > 720$ 1/s (Fig. 3a).

With a further increase in the network load, the condition (11) under $\alpha = \alpha_{TH}$ (Fig. 3a) was provided only by limiting the intensity of the flows that passed through the primary and backup paths. In this case, as in the previous example, Traffic Policing has been differentiated according to the flow priorities and condition (13) as well as according to the packet flow, which provoked network overload and nonfulfillment of the condition (11).

**Table 1.** Order of multipath routing of two flows using the proposed model of traffic engineering fast ReRoute with support of policing (link $E_{11,12}$ protection)

| Link | First flow | | | Second flow | | | $\alpha_{i,j}$ |
|------|--------------|-------------|----------------|--------------|-------------|----------------|------|
|      | Primary path | Backup path | Upper bound | Primary path | Backup path | Upper bound |      |
| $E_{1,2}$ | **788.17** | **900** | **900** | **0** | **0** | **0** | 0.75 |
| $E_{2,3}$ | 592.87 | 675 | 684.45 | 0 | 0 | 0 | 0.74 |
| $E_{1,4}$ | 161.83 | 0 | 302.13 | 0 | 0 | 0 | 0.68 |
| $E_{2,5}$ | 195.30 | 225 | 382.03 | 0 | 0 | 0 | 0.69 |
| $E_{3,14}$ | 406.66 | 449.18 | 470.26 | 0 | 0 | 0 | 0.73 |
| $E_{5,4}$ | **0** | **0** | **0** | **300** | **225** | **300** | **0.75** |
| $E_{5,6}$ | **113.98** | **129.87** | **129.87** | **320.13** | **320.13** | **320.13** | **0,75** |
| $E_{4,7}$ | 161.83 | 0 | 186.00 | 300 | 225 | 326.02 | 0,73 |
| $E_{5,8}$ | **81.32** | **95.13** | **95,13** | **279.87** | **129.87** | **279.87** | **0.75** |
| $E_{6,9}$ | 86.95 | 129.87 | 191.07 | 320.13 | 320.13 | 387.13 | 0.72 |
| $E_{7,11}$ | **0** | **0** | **0** | **189.75** | **225** | **225** | **0.75** |
| $E_{8,9}$ | **52.59** | **95.13** | **95.13** | **96.84** | **129.87** | **129.87** | **0.75** |
| $E_{7,10}$ | 161.83 | 0 | 198.87 | 110.25 | 0 | 153.88 | 0.71 |
| $E_{8,11}$ | 28.73 | 0 | 74.81 | 183.03 | 0 | 207.06 | 0.71 |
| $E_{9,12}$ | 66.14 | 126.71 | 136.75 | 416.97 | 450 | 458.10 | 0.74 |
| $E_{10,11}$ | 161.83 | 0 | 259.70 | 110.25 | 0 | 223.47 | 0,69 |
| $E_{11,12}$ | 190.56 | 0 | 288.73 | 293.28 | 0 | 382.27 | 0,71 |
| $E_{3,13}$ | 186.21 | 225.82 | 245.83 | 0 | 0 | 0 | 0,72 |
| $E_{13,14}$ | 186.21 | 225.82 | 271.31 | 0 | 0 | 0 | 0,71 |
| $E_{6,14}$ | 27.03 | 0 | 169.97 | 0 | 0 | 0 | 0,66 |
| $E_{14,15}$ | **619.90** | **675** | **675** | **0** | **0** | **0** | **0,75** |
| $E_{9,15}$ | 73.40 | 98.29 | 296.10 | 0 | 0 | 0 | 0,67 |
| $E_{15,16}$ | 693.30 | 773.29 | 785.53 | 0 | 0 | 0 | 0,74 |
| $E_{12,16}$ | 256.70 | 126.71 | 403.79 | 0 | 0 | 0 | 0,69 |

As in the previous case (Fig. 2), at the specified output data, the first high-priority packet flow was not limited in its intensity ($\beta^1 = 0$) when using the primary path. The second flow, which had a low priority and was transmitted over the backup path, was constrained with a greater intensity (Fig. 3d). While using the primary path, the second flow was limited with less intensity (Fig. 3c). The first high priority flow was limited only when overloading the links that were a part of the backup path (Fig. 3b).

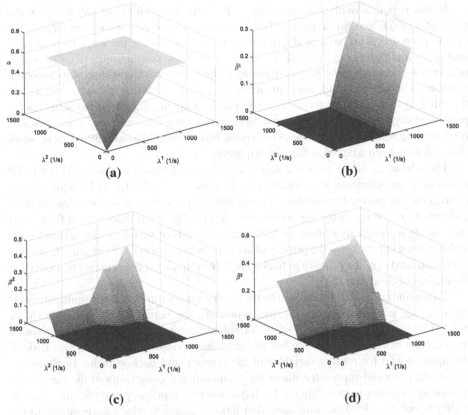

**Fig. 3.** The results of investigation for $\alpha_{TH} = 0.65$.

## 6  Conclusion

It is known that the solutions to increase the fault-tolerance of the communication network with the implementation of fast rerouting are based on the reservation of its elements and bandwidth. That is, the introduction of resource redundancy is a kind of payment for increasing the reliability of network solutions. However, reserving a network resource in the course of implementing a certain protection scheme always adversely affects its performance and the Quality of Service in general. That is, with increasing network load, the implementation of protection scheme of network elements, and especially its bandwidth during fast rerouting, can lead to the overload.

It addition, an effective proactive tool to prevent the network overload is to ensure the balanced use of the available network resource on the principles of the Traffic Engineering concept. An effective reactive method for controlling the overload is Traffic Policing, and in particular the limitation of its intensity at the network edge in accordance with the priority of packet servicing. Therefore, theoretical solutions for ensuring the coordinated solution of network problems with rerouting on the principles of TE and differentiated Traffic Policing are very popular in practice.

In this regard, a mathematical model of fast rerouting with load balancing on the principles of TE and differentiated Traffic Policing in communication networks (1)–(14) is proposed. The basis of the model is the conditions for the implementation of multipath routing (1), and modified conditions for flow conservation (2), (3), which take into account the traffic priority policing at the network edge in case of its probable overload. On the one hand, such an overload can be caused by a load increase. On the other hand, the reason for overloading can be implementation of the schemes for protecting network elements and its bandwidth during the fast rerouting. The conditions for providing protection (redundancy) of a node, link (4–6), and bandwidth of the network (8)–(10) have been adapted to the new requirements.

The advantage of the proposed solution is also the formulation of the TE-FRR-TP problem as an optimization problem with the optimality criterion (12), which firstly focuses on minimizing the dynamically controlled upper bound of communication links utilization ($\alpha$) that meets the requirements of the concept of TE. Secondly, it focuses on minimizing the intensity of flows, which receive denials of service at the network edge and weighted in relation to the priority of service. In the course of prioritizing flows that are transmitted over the primary or backup path (multipath), it is necessary to ensure the fulfillment of condition (13).

The optimization constraints will be the conditions for implementing multipath routing (1), flow conservation (2), (3), and protecting the network link (4), node (6) or bandwidth (8)–(10). The linearity of the formulated optimization problem was provided by a certain extension of the number of calculated variables, (8)–(10), which defined the upper bound for routing variables of the primary and backup paths. Implementation of the proposed approach reduces the computational complexity of the calculation of routing variables responsible for the formation of the primary and backup paths. In addition, balanced load of communication links is ensured, which corresponds to the implementation of the Traffic Engineering concept.

The research of the processes of fast rerouting using the proposed model (1)–(14) on a number of numerical examples (Figs. 1, 2 and 3) implemented in MATLAB environment. Numerical results have confirmed the adequacy and efficiency of the routing solutions made on its basis, both in terms of ensuring their fault-tolerance and load balancing, and for traffic policing. At the same time, in the course of traffic policing, two important principles were realized. First, the limitations concerned primarily the flow that was the source of overload under the condition (11). Second, if the overload created several flows, the limitations particularly concerned a flow with the lower priority in accordance with the conditions (13).

# References

1. Blokdyk, G.: Software-Defined WAN SD-WAN A Clear and Concise Reference. 5STAR-Cooks (2018)
2. Blokdyk, G.: SD-WAN A Complete Guide. 5STARCooks (2018)
3. Naggi, R., Srivastava, R.: SD-WAN The Networking Blueprint for Modern Businesses. Amazon Digital Services LLC (2018)
4. White, M.B.: Computer Networking: The Complete Guide to Understanding Wireless Technology, Network Security, Computer Architecture and Communications Systems (Including Cisco, CCNA and CCENT). CreateSpace Independent Publishing Platform (2018)

5. Monge, A.S., Szarkowicz, K.G.: MPLS in the SDN Era: Interoperable Scenarios to Make Networks Scale to New Services. O'Reilly Media, Sebastopol (2016)
6. Zaitsev, D.A., Shmeleva, T.R., Retschitzegger, W., Pröll, B.: Security of grid structures under disguised traffic attacks. Clust. Comput. **19**(3), 1183–1200 (2016). https://doi.org/10.1007/s10586-016-0582-9
7. Smelyakov, K., Dmitry, P., Vitalii, M., Anastasiya, C.: Investigation of network infrastructure control parameters for effective intellectual analysis. In: Proceedings of the 2018 14th International Conference on Advanced Trends in Radioelecrtronics, Telecommunications and Computer Engineering (TCSET), pp. 983–986. IEEE (2018). https://doi.org/10.1109/tcset.2018.8336359
8. Ruban, I.V., Churyumov, G.I., Tokarev, V.V., Tkachov, V.M.: Provision of survivability of reconfigurable mobile system on exposure to high-power electromagnetic radiation. In: Selected Papers of the XVII International Scientific and Practical Conference on Information Technologies and Security (ITS 2017). CEUR Workshop Processing, pp. 105–111 (2017)
9. Papán, J., Segeč, P., Palúch, P., Mikuš, Ľ., Moravčík, M.: The survey of current IPFRR mechanisms. In: Janech, J., Kostolny, J., Gratkowski, T. (eds.) SDOT 2015. AISC, vol. 511, pp. 229–240. Springer, Cham (2017). https://doi.org/10.1007/978-3-319-46535-7_18
10. Al-shawi, M., Laurent, A.: Designing for Cisco Network Service Architectures (ARCH) Foundation Learning Guide: CCDP ARCH 300-320, 4th edn., Cisco Press, Indianapolis (2017)
11. Mauthe, A., et al.: Disaster-resilient communication networks: principles and best practices. In: Proceedings of the 2016 8th International Workshop on Resilient Networks Design and Modeling (RNDM), pp. 1–10. IEEE (2016). https://doi.org/10.1109/RNDM.2016.7608262
12. Ross, K., Kurose, J.: Computer Networking: A Top-Down Approach, Global Edition, 7th edn. Pearson Higher Education, London (2016)
13. Golani, K., Goswami, K., Bhatt, K., Park, Y.: Fault tolerant traffic engineering in software-defined WAN. In: Proceedings of the 2018 IEEE Symposium on Computers and Communications (ISCC), pp. 01205–01210. IEEE (2018). https://doi.org/10.1109/ISCC.2018.8538606
14. Tomovic, S., Radusinovic, I.: A new traffic engineering approach for QoS provisioning and failure recovery in SDN-based ISP networks. In: Proceedings of the 2018 23rd International Scientific-Professional Conference on Information Technology (IT), pp. 1–4. IEEE (2018). https://doi.org/10.1109/SPIT.2018.8350854
15. Lin, S.C., Wang, P., Luo, M.: Control traffic balancing in software defined networks. Comput. Netw. **106**, 260–271 (2016)
16. Wang, Y., Wang, Z.: Explicit routing algorithms for internet traffic engineering. In: Proceedings of the Eight International Conference on Computer Communications and Networks (Cat. No. 99EX370), pp. 582–588. IEEE (1999). https://doi.org/10.1109/ICCCN.1999.805577
17. Seok, Y., Lee, Y., Kim, C., Choi, Y.: Dynamic constrained multipath routing for MPLS networks. In: Proceedings of the Tenth International Conference on Computer Communications and Networks (Cat. No. 01EX495), pp. 348–353. IEEE (2001). https://doi.org/10.1109/ICCCN.2001.956289
18. Mendiola, A., Astorga, J., Jacob, E., Higuero, M.: A survey on the contributions of software-defined networking to traffic engineering. IEEE Commun. Surv. Tutor. **19**(2), 918–953 (2017). https://doi.org/10.1109/COMST.2016.2633579
19. Prabhavat, S., Nishiyama, H., Ansari, N., Kato, N.: On load distribution over multipath networks. IEEE Commun. Surv. Tutor. **14**(3), 662–680 (2012). https://doi.org/10.1109/SURV.2011.082511.00013
20. Koryachko, V.P., Perepelkin, D.A., Byshov, V.S.: Development and research of improved model of multipath adaptive routing in computer networks with load balancing. Autom. Control. Comput. Sci. **51**(1), 63–73 (2017). https://doi.org/10.3103/S0146411617010047

21. Koryachko, V.P., Perepelkin, D.A., Byshov, V.S.: Enhanced dynamic load balancing algorithm in computer networks with quality of services. Autom. Control. Comput. Sci. **52**(4), 268–282 (2018). https://doi.org/10.3103/S0146411618040077

22. Perepelkin, D., Byshov, V.: Visual design environment of dynamic load balancing in software defined networks. In: Proceedings of the 2017 27th International Conference Radioelektronika (RADIOELEKTRONIKA), pp. 1–4. IEEE (2017). https://doi.org/10.1109/RADIOELEK. 2017.7936643

23. Wang, N., Ho, K., Pavlou, G., Howarth, M.: An overview of routing optimization for internet traffic engineering. IEEE Commun. Surv. Tutor. **10**(1), 36–56 (2008). https://doi.org/10.1109/ COMST.2008.4483669

24. Zhang, X., Cheng, Z., Lin, R., He, L., Yu, S., Luo, H.: Local fast reroute with flow aggregation in software defined networks. IEEE Commun. Lett. **21**(4), 785–788 (2017). https://doi.org/ 10.1109/LCOMM.2016.2638430

25. Malik, A., Aziz, B., Adda, M., Ke, C.H.: Optimisation methods for fast restoration of software-defined networks. IEEE Access **5**, 16111–16123 (2017). https://doi.org/10.1109/ACCESS. 2017.2736949

26. Rzym, G., Wajda, K., Chołda, P.: SDN-based WAN optimization: PCE implementation in multi-domain MPLS networks supported by BGP-LS. Image Process. Commun. **22**(1), 35–48 (2017). https://doi.org/10.1515/ipc-2017-0004

27. Luo, M., Zeng, Y., Li, J., Chou, W.: An adaptive multi-path computation framework for centrally controlled networks. Comput. Netw. **83**, 30–44 (2015). https://doi.org/10.1016/j. comnet.2015.02.004

28. Lemeshko, O.V., Garkusha, S.V., Yeremenko, O.S., Hailan, A.M.: Policy-based QoS management model for multiservice networks. In: Proceedings of the 2015 International Siberian Conference on Control and Communications (SIBCON), pp. 1–4. IEEE (2015). https://doi. org/10.1109/SIBCON.2015.7147124

29. Lemeshko, A.V., Evseeva, O.Y., Garkusha, S.V.: Research on tensor model of multipath routing in telecommunication network with support of service quality by greate number of indices. Telecommun. Radio Eng. **73**(15), 1339–1360 (2014). https://doi.org/10.1615/Teleco mRadEng.v73.i15.30

30. Lemeshko, O., Arous, K., Tariki, N.: Effective solution for scalability and productivity improvement in fault-tolerant routing. In: Proceedings of the 2015 Second International Scientific-Practical Conference Problems of Infocommunications Science and Technology (PIC S&T), pp. 76–78. IEEE (2015). https://doi.org/10.1109/INFOCOMMST.2015.7357274

31. Lemeshko, O.V., Yeremenko, O.S.: Dynamics analysis of multipath QoS-routing tensor model with support of different flows classes. In: Proceedings of the 2016 International Conference on Smart Systems and Technologies (SST), pp. 225–230. IEEE (2016). https://doi.org/10. 1109/SST.2016.7765664

32. Yeremenko, O.S., Lemeshko, O.V., Tariki, N.: Fast ReRoute scalable solution with protection schemes of network elements. In: Proceedings of the 2017 IEEE First Ukraine Conference on Electrical and Computer Engineering (UKRCON), pp. 783–788. IEEE (2017). https://doi. org/10.1109/UKRCON.2017.8100353

33. Lemeshko, O., Yeremenko, O., Hailan, A.M.: Two-level method of fast ReRouting in software-defined networks. In: Proceedings of the 2017 4th International Scientific-Practical Conference Problems of Infocommunications. Science and Technology (PIC S&T), pp. 376–379. IEEE (2017). https://doi.org/10.1109/INFOCOMMST.2017.8246420

34. Lemeshko, O., Yeremenko, O.: Enhanced method of fast re-routing with load balancing in software-defined networks. J. Electr. Eng. **68**(6), 444–454 (2017). https://doi.org/10.1515/jee-2017-0079

35. Lemeshko, O., Yeremenko, O.: Linear optimization model of MPLS traffic engineering fast ReRoute for link, node, and bandwidth protection. In: 2018 14th International Conference on Advanced Trends in Radioelecrtronics, Telecommunications and Computer Engineering (TCSET), pp. 1009–1013. IEEE (2018). https://doi.org/10.1109/TCSET.2018.8336365

# Performance Comparison Between Traditional Network and Broadcast Cognitive Radio Networks Using Idle Probability

Sarab K. Mahmood[✉] and Basma N. Nadhim

Al Mustansiriyah University, Baghdad, Iraq
sarabkamalm@gmail.com, besma.nazar@uomustansiriyah.edu.iq

**Abstract.** In this paper, we suggest a broadcast routing protocol for multi-hop mobile Ad Hoc CRN based on the connotation of the shortest path tree (SPT). The suggested algorithm uses the probability of success (POS) metric in executing the channel customize process. Under different network parameters within Rayleigh fading channels, the performance of SPT broadcast CRN with POS scheme is compared with the performance of maximum data rate (MDR), maximum average spectrum availability (MASA) and random channel assignment schemes. After simulation, SPT broadcast CRN with POS scheme was seen to pretend the best execution versus other schemes. The network's performance improved with the progressing in the value of idle probability ($P_I$) because the primary user's (PU) traffic load is low when the value of $P_I$ is high. In addition, the performance of SPT broadcast CRN is compared with the performance of traditional network and notice that it is nearing the latter is performance as the value of $P_I$ increases.

**Keywords:** Broadcast · CRN · POS

## 1 Introduction

### 1.1 Cognitive Radio

A cognitive radio is defined as a proportion that can shift its sender parameters according to the climate in which it works [1]. The capability of a cognitive radio (CR) is the emotion and collect information from the surrounding climate as well as has the capability to readily adjust the parameters of the operation, for optimum execution, according to the data perceived [2]. The cognitive radio technology is achieved as the opener stabilization technology for the following generation dynamic spectrum arrival networks that can soundly use the obtainable low usage shade appropriated by the Federal Communications Commission (FCC) to authorized possessors, known as primary users. Cognitive radios become strong the employment of tentatively not used shade, indicate to as white space or spectrum hole [2] and the secondary user should fluently move to different shade hole or biding in the selfsame band when a primary user proposes to use this band, modifying its transition power standard or modulation sketch to avert overlapping with the primary user Cognitive radios simplify a more resilient and inclusive use of the bounded and low

© Springer Nature Switzerland AG 2020
A. M. Al-Bakry et al. (Eds.): NTICT 2020, CCIS 1183, pp. 44–58, 2020.
https://doi.org/10.1007/978-3-030-55340-1_4

usage spectrum for the secondary users, who have no spectrum licenses. In Traditional shade distribution sketches when secondary employers are permitted to send information along with primary employers, the transitions should not overlap with all other beyond a sill. On the other side, the secondary employers can sending the data only in the obscurity of primary employers, and the secondary employer must be have the ability to expose the reoccurrence of the primary employer and lift the band [3].

The procedure of employing the spectrum holes by cognitive user when the primary user is not obtainable and reversions the channel again when it is wanted by the primary user is referred as the Dynamic Spectrum Access (DSA) as in Fig. 1. The topical of the cognitive network is to use the obtainable idle spectrum soundly without causing over lap to the primary users [4].

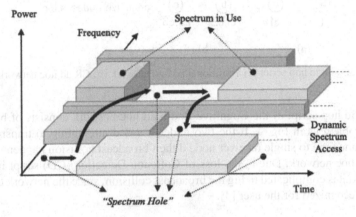

**Fig. 1.** Principle of Cognitive Radio network.

## 1.2 Broadcast Channel

In wireless ad hoc network control the information is broadcasted to purpose node through moderate nodes. Cognitive Radio technologies for Broadcasting have began to catch the scientific research recently. Broadcasting is the better technicality to communicate security messages to the network. Broadcast channel has one transmitter and many receivers [5]. For example:

A broadcast channel consists of one input $x$ and two output $y_1$ and $y_2$ and a probability transition function $p(y_1, y_2|x)$ as in Fig. 2 [6]:

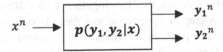

**Fig. 2.** Broadcast channel.

In traditional ad hoc network as in Fig. 3a, broadcast is accomplished by using joint channel due to the existence of single spectrum. In Cognitive Radio (CR) ad hoc network as in Fig. 3b, PU have authorized channels, if the job is finished by the primary user within a time period, then the channels are release for the use of secondary user. In single hop script the broadcasting at single time aperture is complicated due to the existence of complicated channels. So it is not contingent for the adjoining nodes to receive data at selfsame time [4].

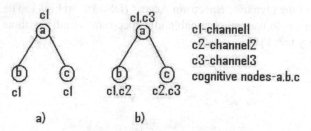

**Fig. 3.** Single hop script (a) Traditional ad hoc network (b) CR ad hoc network.

The multi hop script of the Cognitive Radio ad hoc network consists of broadcast collision problem as in Fig. 4. If the two nodes c and d attempting to transmit same packet at same time to single receiver node d then broadcast collision happens in multi hop CR ad hoc network. Due to the loss of Collision Detection (CD) script in CR ad hoc network, it is complicated to lug the broadcast collision, since the network topology data is not recognized for the user [4].

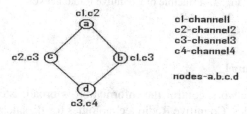

**Fig. 4.** Multi hop CR network.

## 2  System Paradigm

We deem a cognitive radio network that live together with different primary radio networks in the selfsame geographical zone, where there is one exporter trying to send information to Nr receivers through a combination of obtainable channels C as shown in Fig. 5. The case of each channel is shaped as a two cases Markov model, rotated between occupied and idle. Occupied case acts that a primary users conquers the channel, so, the secondary users cannot utilize this channel. The idle case acts that the primary users do not utilize the channel, so, the secondary users can opportunistically utilize this channel. We suppose infrastructure less ad hoc multi-hop cognitive radio network. We suppose that a joint control channel is obtainable to assortment the sending in the cognitive radio network [7].

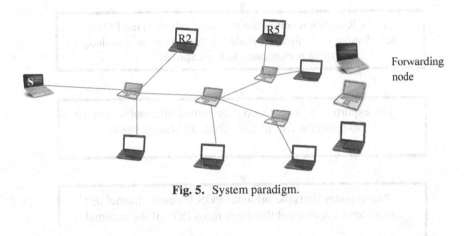

**Fig. 5.** System paradigm.

# 3   Dissection of System Model Using Probability of Success

We assume a multi-layer on-demand broadcast routing protocol over multi-hop cognitive radio networks. The system model have a primary users all with the selfsame bandwidth symbolized by BW. We deem N cognitive radio (CR) users. Every cognitive radio (CR) employer j (where $N \geq j \geq 1$) can singly feeling and discover idle channels. Every CR employer is armed with two half-duplex radio transmitter-receiver that can at the same time be utilized. One transmitter-receiver is meshed to a common control channel (CCC), while the other can be dynamically meshed to any obtainable data channel, but at a fixed time, it can only work over one information channel. The data transmitter-receiver is a recurrence smart [8]. We suppose that adjoining CR employers have an careful data about the goodness of channels and their rate shade-availability time. This data can be prophesied from the statistical history of PU actions. The sketch that selects the channel with maximum average spectrum availability (MASA) does not pick into account the goodness of that channel on account of channel availability does not be inverted the channel goodness. On the other side, the sketch that selects the channel with maximum data rate (MDR) does not deem the availability time of that channel. The probability of success POS measure together deems the actions of both channel goodness and its availability time [9].

The following sketch epitomize the procedures of the channel assignment operation using POS metric (Fig. 6):

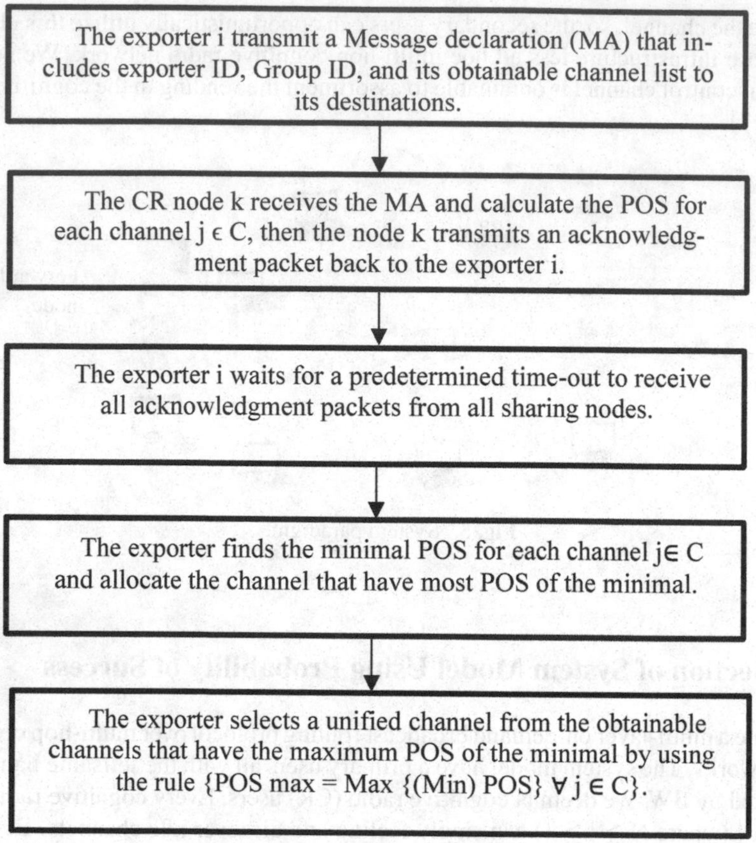

The exporter i transmit a Message declaration (MA) that includes exporter ID, Group ID, and its obtainable channel list to its destinations.

The CR node k receives the MA and calculate the POS for each channel j ϵ C, then the node k transmits an acknowledgment packet back to the exporter i.

The exporter i waits for a predetermined time-out to receive all acknowledgment packets from all sharing nodes.

The exporter finds the minimal POS for each channel j∈ C and allocate the channel that have most POS of the minimal.

The exporter selects a unified channel from the obtainable channels that have the maximum POS of the minimal by using the rule {POS max = Max {(Min) POS} ∀ j ∈ C}.

Fig. 6. A simplified POS algorithm.

The probability of success POS between any two nodes $i$ and $k$ through channel $j \in$ C based on a stochastic paradigm of PUs actions beneath a Rayleigh fading channel paradigm can be calculated as in Eq. (1):

$$Pr_{(j)}^{(i-k)} = \frac{p_t}{d^n} \left( \frac{\lambda}{4\pi} \right)^2 \left( \xi_{(j)}^{(i-k)} \right) \tag{1}$$

where $Pr_{(j)}^{(i-k)}$ explains the received power from sender $i$ to recipient $k$
$p_t$ is the cognitive radio transition power
$d$ refer to the distance between any two nodes
$n$ refer to the path loss exponent
and $\xi_{(j)}^{(i-k)}$ is the channel power gain between nodes $i$ and $k$ over channel $j$.

$$R_{(j)}^{(i-k)} = (Bw)log_2 \left( 1 + \frac{Pr_{(j)}^{(i-k)}}{Bw * N_0} \right) \tag{2}$$

where $R_{(j)}^{(i-k)}$ is the data rate between nodes $i$ and $k$ over channel j
$N_0$ explains the thermal power spectral density in (Watt/Hz)
and $Bw$ is the channel bandwidth for channel $j$.

$$Tr_{(j)}^{(i-k)} = \frac{D}{R_{(j)}^{(i-k)}} \tag{3}$$

Where $Tr_{(j)}^{(i-k)}$ is the wanted sending time to transmit a packet from node $i$ to $k$ through channel j
And $D$ is the packet size (in bits).
Finally:

$$P_{suc(j)}^{i-k} = exp\left(\frac{-Tr_{(j)}^{(i-k)}}{\mu_j}\right) \tag{4}$$

where: $P_{suc(j)}^{i-k}$ is the probability of success POS between any two nodes $i$ and $k$ through channel $j$
$\mu_j$ refer to the average spectrum availability time for channel $j$.

The expense of utilizing probability of success POS as a channel task sketch is to define the rate obtainable time of the channel and to determine the goodness of that channel. While the expense of utilizing maximum average shade availability sketch or maximum data rate sketch demands only limitation the channel goodness or learning the average availability time of the channel, respectively.

is $(n \log n + m)$, where $n$ is the number of vertices and $m$ is the number of edges that connect the nodes [10, 11].

The major concept of the sketch is to send data from a specified exporter to multi receivers in a multi-hop CRN using Shortest Path Tree (SPT) such that the network throughput is refined. The SPT gets it the loop free path with minimal distance. As shown in Fig. 7 (for any linked and undirected graph with a single exporter), the SPT can be found utilizing Dijkstra's algorithm as in Fig. 8. In order to build the tree, the SPT suppose an presence of a exporter node, so that the whole tree should be changed when the exporter node is changed. The whole running time for Dijkstra's algorithm is $(n \log n + m)$, where $n$ is the number of vertices and $m$ is the number of edges that connect the nodes [10, 11].

At first, the sketch transforms the network topology to a tree steadfast at the exporter and extending all the nodes. A multi-layer transition is needed to transmit information packets form the exporter to all receivers in the tree because of the multi-hop feature in the CRN.

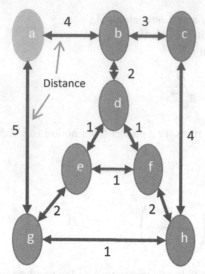

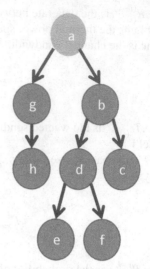

**Fig. 7.** Linked and undirected topology with $m = 11$ and $n = 8$.

**Fig. 8.** The produced shortest path tree.

## 4    Simulation Results

At first, we make a random topology with one CR exporter and N cognitive radio destination within zone. We deem a Rayleigh fading channel paradigm with path-loss exponent (n) to characterize the profit between any two connected nodes in the network [12]. we put the thermal-noise power spectral density $N_0 = 10^{-18}$ W/Hz and bandwidth BW = 1 MHz for all channels. The transition power and information packet size is put to $p_t = 0.4$ Watt D = 4 KB respectively.. The average availability time for each PR channel of PRNs is modeled as an Idle/Busy Markov model with average ON and OFF duration. The whole number of primary channels (M) is 30 with average shade availability time ($\mu_j$) that ranges from 2 ms to 70 ms. We deem three states for idle probability (i.e., $p_I = 0.9$ (low PR traffic load)), $p_I = 0.6$ (temperatePR traffic load), and $p_I = 0.3$ (high PR traffic load)). We assume there are sentinel band channel and limited channel occupancy.

## 5    Performance Evaluation for Shortest Path Tree (SPT)

Depending on the SPT method, the suggested algorithm gets the minimal distance from the exporter to the each destination in the network. The shortest path tree is performed in Matlab using Djikestra algorithm [13]. we lead simulations to estimate the performance of the assumed sketch with the state-of-the-art schemes. Then, we estimate the performance of the assumed sketch under variant network parameters. In this collection of experiment, we deem the total number of nodes (N) is 60.

### 5.1 Performance Evaluation of Multicast CRN Under the Effect of Channel Bandwidth

The throughput and packet delivery rate performance for broadcast CRN verses channel bandwidth with three values of Idle probability ($P_I = 0.3, 0.6$, and $0.9$) and for traditional network (i.e. Idle probability $P_I = 1$) are shown in Figs. 9, 10, 11 and 12 respectively. Note that as the channel BW increases the throughput and packet delivery rate increase. This is because the channel capacity (i.e., data rate) is proportional to the channel BW. However, from Fig. 9a and Fig. 10a notice that for small value of Idle probability ($P_I = 0.3$) the channels at most times are busy because of the high primary user traffic load and under this value of $P_I$ the throughput performance of POS at (BW = 4 Hz) is comparable of MASA scheme by 0.1 Mbps and outperforms MDR and RS schemes by (1, 1.2) Mbps respectively and in term of PDR by (0.01, 0.22 and 0.29) packed respectively.

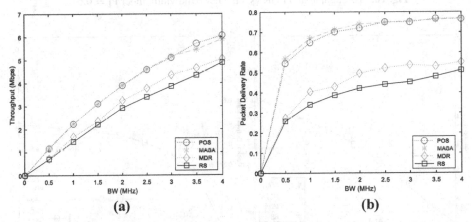

**(a)**                                              **(b)**

**Fig. 9.** Throughput and PDR vs. channel bandwidth under $P_I = 0.3$.

From Fig. 10 and 11 one can notice that as the Idle probability increased ($P_I = 0.6$ and $0.9$) the throughput performance of POS scheme is improved. At $P_I = 0.6$ the performance of POS scheme (at BW = 4 Hz) outperforms MASA, MDR and SR by (0.4, 1.4 and 1.9) Mbps and by (0.7, 1.5, 2) Mbps at $P_I = 0.9$, while for PDR performance the POS scheme is outperform at $P_I = 0.6$ by (0.03, 0.25, 0.35) packed and at $P_I = 0.9$ by (0.04, 0.27 and 0.36) packed respectively.

Figure 12 shows that, the performance of POS scheme for traditional network (at BW = 4 Hz) is outperform MASA, MDR and SR by (0.8, 1.6 and 2.1) Mbps, while for PDR performance the POS scheme is outperform by (0.06, 0.29 and 0.37) packed. This is because that when $P_I$ value increased the more probability that suitable channels are available for CR users transmission because the PU traffic load is low.

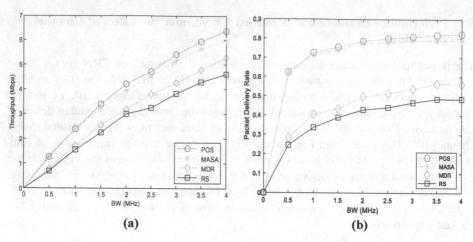

**Fig. 10.** Throughput and PDR vs. channel bandwidth under $P_I = 0.6$.

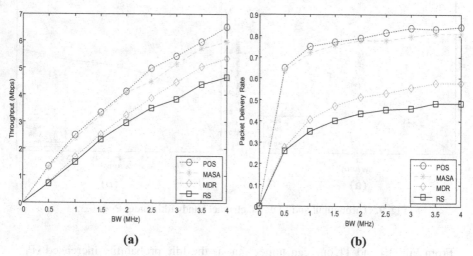

**Fig. 11.** Throughput and PDR vs. channel bandwidth under $P_I = 0.9$.

## 5.2 Performance Evaluation of CRN Under the Effect of Packet Size D

The throughput and PDR performance of multilayer multi hop broadcast CRN verses packed size with three values of Idle probability ($P_I = 0.3, 0.6,$ and $0.9$) and for traditional network (i.e. Idle probability $P_I = 1$) are shown in Figs. 13, 14, 15 and 16 respectively. From all these figures one can notice that the throughput and PDR performance are decreased as the value of packed size D increased, this is because that the

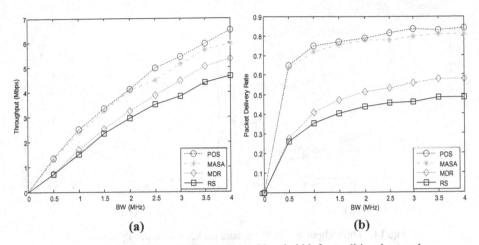

**Fig. 12.** Throughput and PDR vs. channel bandwidth for traditional network.

high value of D need high time available of the channel to transmit the packet date of CR user effectively and this make the chance to find best channel difficult. Figure 13 shows that the improved throughput gains for POS scheme (at D = 10 KB) compare to MASA, MDR and RS schemes at $P_I$ = 0.3 are (0.2, 0.82, 0.97) Mbps and in term of PDR by (0.05, 0.25 and 0.27) packed respectively.

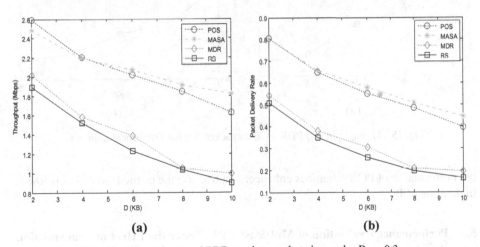

**Fig. 13.** Throughput and PDR vs. data packet size under $P_I$ = 0.3.

Also, Figs. 14, 15 and 16 shows that at $P_I$ = 0.6, 0.9 and for traditional network the improved throughput gains for POS scheme compare to MASA, MDR and RS schemes are (0.01, 0.86, 0.99) Mbps, (0.02, 1, 1.01) Mbps and (0.02, 1.01, 1.02) Mbps respectively and in terms of PDR by (0.01, 0.3 and 0.32) packed, (0.01, 0.33, 0.35) packed and (0.02, 0.33, 0.36) packed respectively. Also, It should be noted that as the value of $P_I$ is increased

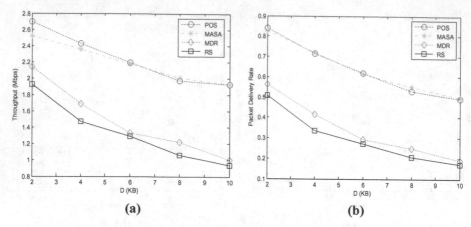

**Fig. 14.** Throughput and PDR vs. data packet size under $P_I = 0.6$.

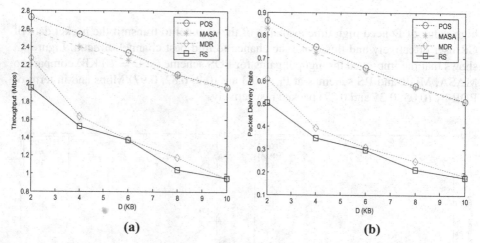

**Fig. 15.** Throughput and PDR vs. data packet size for Traditional network.

the performance of POS scheme is enhancement because the traffic load of PU is low at high value of $P_I$.

### 5.3 Performance Evaluation of Multicast CRN Under the Effect of Transmission Power

The throughput and PDR performance of multilayer multi hop broadcast CRN verses transmission power with three values of Idle probability ($P_I = 0.3, 0.6$, and $0.9$) and for traditional network (i.e. Idle probability $P_I = 1$) are shown in Figs. 17, 18, 19 and 20 respectively. As the transmission power increased the throughput and PDR performance are improved because the time required to transmit data decreased therefore more data can be transmitted over each channel. Also, for high value of Idle probability the

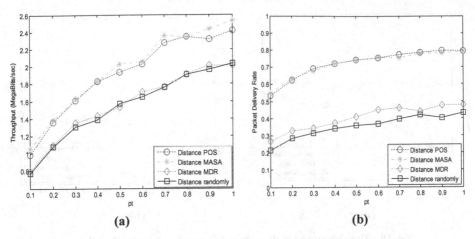

**(a)**                    **(b)**

**Fig. 16.** Throughput and PDR vs. transmission power under $P_I = 0.3$.

performance of POS schemes improved as mention before in previous sections. The improved throughput gains for POS scheme (at $p_t = 1$) compare to MASA, MDR and RS schemes at ($P_I = 0.3, 0.6, 0.9$) and for traditional network are (0.1, 0.5, 0.5) Mbps, (0.15, 1, 1.15) Mbps, (0.2, 1.1, 1.3) Mbps and (0.3, 1.2, 1.4) Mbps respectively. Also, The improved PDR gains for POS scheme compare to MASA, MDR and RS schemes at ($P_I = 0.3, 0.6, 0.9$) and for traditional network are (0.01, 0.29, 0.35) packed, (0.012, 0.3, 0.36) packed, (0.02, 0.3, 0.38) packed and (0.021, 0.34, 0.42) packed respectively.

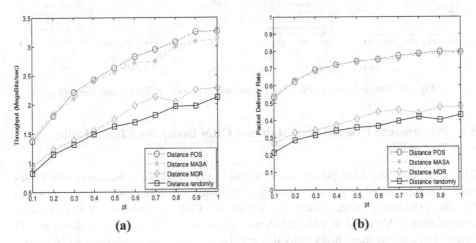

**(a)**                    **(b)**

**Fig. 17.** Throughput and PDR vs. transmission power under $P_I = 0.6$.

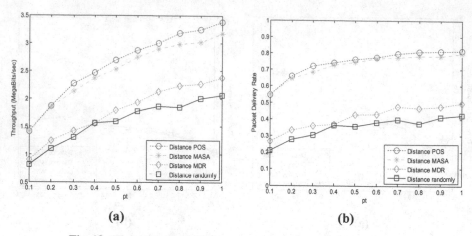

**Fig. 18.** Throughput and PDR vs. transmission power under $P_I = 0.9$.

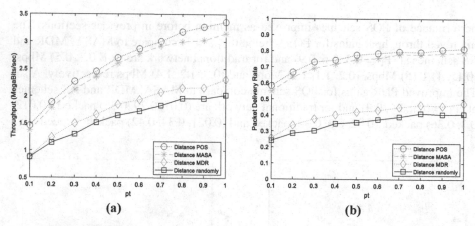

**Fig. 19.** Throughput and PDR vs. Transmission power for Traditional network.

## 5.4 Performance Evaluation of Multicast CRN Under the Effect of Idle Probability

The throughput and PDR performance of multilayer multi hop broadcast CRN verses values of Idle probability is shown in Fig. 20. From this figure one can notice that as the Idle probability increased the throughput and PDR performance of POS scheme outperforms MASA, MDR and SR schemes respectively this is because that when $P_I$ value increased the more probability that suitable channels are available for CR users.

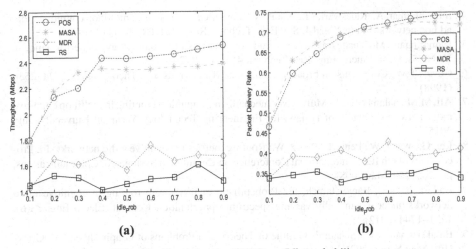

**Fig. 20.** Throughput and PDR vs. Idle probability.

## 6 Conclusion

In this work, SPT is used to construct the broadcast router protocol for mobile Ad Hoc CRN that has multilayer and multi-hop consideration. Under different network parameters the performance of the CRN in terms of throughput and PDR with POS channel assignment is compared with Maximum Average Spectrum Availability (MASA), Maximum Data Rate (MDR) and Random channel assignment schemes within Rayleigh fading channel model consideration and three idle probability (i.e., $P_I = 0.3, 0.6, 0.9$) cases. The simulation results proved that the CRN with POS scheme have the best performance than using other schemes and as the value of $P_I$ increased the performance of POS scheme improved because the chance of finding suitable available channels for CR user's transmission increased because the traffic load of PU is low at high value of $P_I$. In addition, the execution of the CRN is compared with the execution of traditional network in terms of throughput and PDR with POS channel assignment. From these results, we observe that the performance of the SPT is influenced by the value of idle probability ($P_I$) and its performance is approaching the performance of the traditional network as the value of $P_I$ increases, but cannot be equal because some channels will be busy with the primary user (Pu) at variance the traditional network in which all the channels are ready for using by the Pu.

## References

1. Haykin, S., Thomson, D.J., Reed, J.H.: Spectrum sensing for cognitive radio. Proc. IEEE **97**(5), 849–877 (2009)
2. Hykin, S.: Cognitive radio: brain-empowered wireless communications. IEEE J. Sel. Areas Commun. **23**(2), 201–220 (2005)
3. Mahajan, R., Bagai, D.: Cognitive radio technology: introduction and its applications. Int. J. Eng. Res. Dev. **12**(9), 17–24 (2016)

4. Sureshkumar, A., Rajeswari, M.: A survey on broadcast protocols in multihop cognitive radio ad hoc network. International J. Sci. Eng. Technol. Res. (IJSETR) 3(11) (2014)
5. Ramalingam, M., Thangarajan, R.: A study of broadcasting protocols and its performance in VANETs. Int. J. Emerg. Eng. Res. Technol. 4(3), 1–10 (2016)
6. Cover, T.M.: Comments on broadcast channels. IEEE Trans. Inf. Theory 44(6), 2524–2530 (1998)
7. Ali, M.M., Salameh, H.B.: Multi-layer mechanism for multicast routing in multihop cognitive radio networks. Faculty of Hijjawi for Engineering Technology Yarmouk University, July 2015
8. Lei, G., Wang, W., Peng, T., Wang, W.: Routing metrics in cognitive radio networks. In: Proceedings of 4th IEEE International Conference on Circuits and Systems for Communications, pp. 265–269, May 2008
9. Badarneh, O.S., BanySalameh, H.: Probabilistic quality aware routing in cognitive radio networks under dynamically varying spectrum opportunities. Comput. Electr. Eng. 38(6), 1731–1744 (2012)
10. Balakrishnan, V.K.: Schaum's Outline of Theory and Problems of Graph Theory. McGraw-Hill, New York (1997)
11. Wu, B.Y., Chao, K.: Spanning Trees and Optimization Problems. CRC Press, Boca Raton (2004)
12. Mhaidat, Y., et al.: A cross-layer video multicasting routing protocol for cognitive radio networks. In: Seventh International Workshop on Selected Topics in Mobile and Wireless computing (2014)
13. MathWorks. http://www.mathworks.com

# A Developed Cloud Security Cryptosystem Using Minhash Technique

Mehdi Ebady Manaa$^{(\boxtimes)}$ and Rasha Hussein Jwdha

College of Information Technology, University of Babylon, Babylon, Iraq
meh_man12@yahoo.com,
{mehdi.ebady,rashahussein}@itnet.uobabylon.edu.iq

**Abstract.** Data protection becomes an important issue for companies and organizations, especially for files uploaded in a place that can be accessed from anywhere and at any time such as cloud computing, which enables multiple users to access information through the provider by the cloud leased service. This access has enabled many unauthorized people to jeopardize files, and is considered a persistent problem at present. Thus, the increased demands for the adoption of cloud computing technologies has led to an increased focus on data protection. In this paper, cloud computing using Salesforce.com is used to access and upload the encrypted files using a hybrid cryptosystem from Advanced Encryption Standard (AES) and Blowfish algorithms. The implemented approach has used the principle of the Minhash technique to generate a group of keys in an efficient way using the k-shingle and Hash function.

The implemented work is evaluated using the following performance criteria: encryption time, throughput, memory usage, avalanche effect and entropy by uploaded encrypted files to the cloud using Object Access Protocol- Application Programming Interface (SOAP API) with force.com web service connector cloud. The obtained results have shown a promising way to protect organizations' sensitive files in less encryption time and high randomness. A hybrid cryptosystem of Blowfish and AES shows a better performance, in terms of encryption time and security (for 1 MB is 60.1 s), throughput (21249.28 bytes/sec), memory used (638561 KB) and avalanche effect (98%).

**Keywords:** Cloud computing · Hybrid encryption minhash algorithm · K-shingle · SOAP

## 1 Introduction

Cryptography uses security criteria represented by Integrity, Availability, and Confidentiality (CIA) to protect data for the large organizations when accessesing the Internet. It also provides authentication and non-repudiation for these organizations.

It is necessary to pay attention to data security and methods of protecting data stored in computers and communication networks devices. Information security methods provide defense against attacks and unauthorized access, use, disclosure, disruption, modification, or destruction [1]. The modern cipher categories can be classified into the following sub categories: -

© Springer Nature Switzerland AG 2020
A. M. Al-Bakry et al. (Eds.): NTICT 2020, CCIS 1183, pp. 59–72, 2020.
https://doi.org/10.1007/978-3-030-55340-1_5

The shared key is used in the symmetric algorithms for both the encryption and decryption process. The main advantages of the symmetric algorithms involve their easy and fast implementation while their main drawbacks are key implementation and management and limited distribution security. The Data Encryption Standard (DES), Triple Data Encryption Standard (3DES), and International Data Encryption Algorithm (IDEA), Advanced Encryption Standard (AES) and Blowfish are examples of shared key encryption algorithms [2, 3].

The asymmetric key encryption (public key encryption) is a type of encryption where the user has a pair of encryption keys, a public key and a private key. The private key remains secret (special for the person who will decrypt the message). The public key is shared between the parties and with everyone. The advantage of this type is that when the message is encrypted using the public key, it can only be decrypted by the corresponding private key. The main disadvantages of this type of encryption is that it is slower than that which uses the symmetric algorithm. Examples of asymmetric algorithms are Rivest-Shamir-Adleman (RSA), Digital Signature Algorithm (DSA), Diffie Hellman (DH) [4].

Cloud computing is a new technology used by large organizations to access and share their files. It utilizes the principle of multitenancy by which many users can access and share the same resources in the cloud. The main definition of the cloud computing can be found in [5]. At the same time, the main concern in cloud computing is to provide a secure platform to access the files for these organizations. The cloud security is considered a main challenge in the research work [6]. The security is an obstacle to accelerated adoption and deployment of cloud computing for organizations and governments [7].

The organizations have a large amount of sensitive data that needs to be protected from unauthorized access. It is noticed by the large cloud providers that data security has been of great importance in the recent research work [6, 8, 9]. Therefore, the security in the cloud is an open challenge for the organizations.

In this paper, a proactive and robust approach is presented to protect the essential information of an organization. The implemented approach uses the block cipher algorithm based on a k-shingle with Minhash technique. It introduces a new developed method to generate keys. In addition, the hybrid cryptography method to encrypt and decrypt the files that include sensitive files of the organizations is also designed and implemented to upload the files to the cloud platforms (Salesforce.com [10]) using Force.com SOAP AP to address data security problems in the cloud computing.

## 2  Related Works

This section intends to shed light on some studies relevant to the present one. Many researchers use symmetric algorithms to encrypt the sensitive data and then they use the same key to decrypt it to the original form.

A new hybrid method in [11], merges two types of cryptography algorithms: the symmetric algorithm with the asymmetric ones, i.e. the hybrid system between RSA and AES algorithms. This method provides a secure approach for the use of data in the cloud environment. The basic implementation of this work, was to provide the generated keys on the basis of system time which prevented users or even administrator from knowing the keys in the cloud environment. There were three different keys (private key, public keys, and secret keys) that increased the strength of the encryption system. Data encryption can only be decrypted using the private and secret keys, known by the user only. The study in [12] applied the DES algorithm, the AES algorithm and the hybrid AES-RSA algorithm are used to encrypt and decrypt the data. The proposed method was suggested to merge the AES and RSA algorithms and to increase the length of the RSA key. The hybrid method was implemented to encrypt the lightweight data in the cloud storage environment in order to enhance the confidentiality of data in the cloud environment. A hybrid system using Blowfish and MD5 is presented in [13]. The authors compared Blowfish and MD5 to RSA with MD5 in terms of time and storage values. The results show less time in the case of the Blowfish and MD5 comparing with RSA and MD5.

The proposed study in [14], a hybrid cryptosystem to achieve confidentiality, authentication and integrity, is conducted by applying the Blowfish symmetric algorithm for encryption and RSA asymmetric algorithm to achieve the confidentiality. This work deals with authentication and the secure hash algorithm for the integrity and to provide a hybrid algorithm on a high-level security for data transmission over the Internet. It also provides a convenient network access to the cloud computing resources and storage applications. A hybrid method in [15] is proposed by merging two types of cryptography AES and Blowfish algorithms to secure patients' medical report data into the PDF form and generate crypto keys used in the two methods (i.e. RSA and ECC methods). The results show that the time and throughput for the Elliptic Curve Cryptography (ECC) has a higher strength perbit and smaller key size and a lower memory.

A hybrid technique for cryptography using two algorithms (AES and Blowfish) is introduced in [16]. The MD5 was used to hash the key which will increase the security of the key and increase security using a hybrid cryptography. The evaluation metrics are based on encryption time, decryption time, and throughput. In addition, the study in [17] proposed a method to fragment the encryption of the documents and to generate keys by variable length based on the degree of confidentiality for each fragment encryption. The results show an optimal solution in the case of medium length documents and a low degree of confidentiality. The AES encryption algorithm using Heroku cloud platform is presented in [18]. The evaluation based on delay calculation for data encryption shows that big data leads to increase of the delay time for the encrypting data.

## 3  Cloud Security

The main challenges of the cloud security are virtualization and multi-tenancy for the data in the cloud computing [19]. These challenges are classified as the main security challenges for the cloud provider and cloud user. There are many types of violation to the cloud platforms to access sensitive data. Thus, the solutions are to use cryptography algorithms, authentication scheme, data encryption, access control, and authorization [20]. It is necessary to encrypt the data before uploading them to the cloud computing because there are many types of violation access either from inside or outside the organizations. It is necessary to provide the cloud secure approach for the data uploaded in the cloud. In this paper, many encryption algorithms are implemented to make a robust cryptosystem. In addition, a hybrid approach of more than one encryption algorithm is applied. The key generation becomes an input for the other algorithm. Salesforce.com Cloud Computing Salesforce is a cloud platform that uses the software and developing environment for the application. It is used in sales marketing and has an account for developing services. It uses Apex programming language as an object-oriented programming and the visual force as a tag-based markup language for developing applications [21]. The services in Salesforce are classified into three types:

- **Salesforce as SaaS:** no installations, setup or download are required. Log in and use the software across the cloud.
- **Salesforce as PaaS:** no separate platform is required. Develop application-based coding on Salesforce platform to create and deploy application.
- **Salesforce as IaaS:** no hardware or installations are required. All data and applications are stored and they can be secured on Salesforce cloud.

## 4   Hybrid Encryption Algorithms

Hybrid encryption merges two or more types of cryptography algorithms such as AES with Blowfish. Advanced Encryption Standard (AES) is one of symmetric key block cipher and the block size is 128 bits. It has three key size 128, 192 and 256 bits. The AES rounds are 10 when key size 128, 12 and 14 rounds with key size 192 and 256, respectively. AES uses a key expansion to encrypt/decrypt data where that key comes before each of encryption and decryption processes [22].

The AES round key and rounds for many stages for the encryption and decryption are shown in [23]. While Blowfish is symmetric key block cipher algorithm with block size 64 bits and variable key cipher length (32 to 448 bits) to protect the encrypted data. The structure of this algorithm is the fiestal network.

## 5   The Proposed Cloud-Based Cryptosystem

The cloud-based cryptosystem consists of many steps as follows: -

### 5.1   Key Generated Proposed Phase: Minhash Technique

The Minhash is mainly used for the large-scale data in different applications. It is used in the document similarity as one of the Google techniques [24]. It consists of the following main steps: -

**K-shingle (N-gram)**
The term of K-shingle is used heavily in document similarity. In this technique, documents are split into a set of tokens depending on the length of k. For example, let the document have the string "The weather is nice and the sun is shining". If we choose k = 3, the number of the generated tokens equals $(n - k + 1)$ where n is the total number of words in documents and k is the shingle length, as in Table 1.

In this paper, we split the text files into shingles based on the number length of k. Then we apply the Hash functions (MD5) for each shingle. The Minhash function is applied for the k hashed tokens.

**Minhash Function**

The Minhash function converts tokenized text into a set of hash integers, then selects the minimum value. The general form of the Minhash function is given in Eq. (1) [25]

$$h(x) = ax + b \bmod P \tag{1}$$

where a and b are two random values, x is the Hash function value for the tokens and $P$ is the prime number which is greater than the maximum number of x [26]. The principle of the Minhash technique is applied in this work and the Minhash steps are used to generate the keys using one hash function, as shown in Table 1.

Table 1 clarifies the generation of the group keys in details when the k value of k-shingle equals 3. Suppose the hash MD5 output is h, Minhash output is min, and the key generated is k.

**Table 1.** Keys generation steps at k = 3

| #Shingling | #Hash | #Minhash | #Keys |
|---|---|---|---|
| The weather is | h1 | min1 | k1 |
| weather is nice | h2 | min2 | k2 |
| is nice and | h3 | min3 | k3 |
| nice and the | h4 | min4 | k4 |
| and the sun | h5 | min5 | k5 |
| the sun is | h6 | min6 | k6 |
| sun is shining | h7 | min7 | k7 |

Algorithm (1) shows the main steps for generating a group of keys based on Minhash technique.

| Algorithm (1): The Minhash Steps for Generating Cryptosystem Keys |
|---|
| Input:D (Input Data) |
| Output: $Ki = \{K1, ...., Kn\}: K_i \subset h(x_i) = (ax_i + b \bmod P)$ (Group of Keys from K-shingles) |
|   Begin |
| 1.   Apply k-shingle on D |
| 2.    Set K value for K-shingle |
| 3.    Calculate Function k-shingle (k, D) |

```
Algorithm (1): The Minhash Steps for Generating Cryp-
tosystem Keys
4.      Read (D)
5.      Word[] ← Preprocessing (D)
6.      shingles ← splitWord[] into shingles based on(n-k
        +1)
7.      return shingles
8.    end K-Shingle
/* Shingles ← return from applied function k-shingle
9.    Apply MD5 Hash function
10.     Calculate hash MD5(Shingles)
11.       Foreach shingle in k-shingle (k, D)
12.       Hash value ← Hashing shingles (MD5(shingles))
13.       End foreach
14.       x ← Convert Hash value to number
15.       returnx
16.     end hash MD5
        /* X ← return numbers fromMD5 output
17.   Apply Minhash Equation (h(xᵢ)← (axᵢ + b) mod P )
18.   Calculate Minhash Function
19.   Repeat
20.   Foreachx in hash MD5 (Shingles)
21.    for i ← 1 to 10 do
22.    Random rand ← new Random();//define class of random
23.    a = rand.nextint(100);     // range (0- 100)
24.     b = rand.nextInt(100);    // range (0- 100)
25.       If(P > x &&P > a &&P >
          b && P is checked prime)then//a,b, P    isparame-
          ters of Minhash Equation(1)
26.         h(xᵢ) ← (axᵢ + b mod P)
27.         ListMinhash.add(h(xᵢ))
28.        else
29.        return (go to step 20)
30.        end if
31.     end for
32.   return min (ListMinhash)//min is minhash value
33.    end for each
34.   Apply Hash function again
35.   Ki ←MD5 (min (ListMinhash))
36.   return Ki
37.   Until (Minhash calculate of each shingle)
38. End Minhash Function
End Algorithm
```

Here line 17: $h(x_i) \leftarrow (ax_i + b)\, mod\, P$; line 25: $If(P > x\,\&\&P > a\,\&\&P > b\,\&\&\,P\ is\ checked\ prime)then$; line 26: $h(x_i) \leftarrow (ax_i + b\ mod\ P)$; line 27: $ListMinhash.add(h(x_i))$; line 32: $return\ min\,(ListMinhash)$; line 35: $K_i \leftarrow MD5\,(min\,(ListMinhash))$; line 36: $return\ K_i$.

## 5.2  The Hybrid Encryption Phase

In this phase, the k value of k-shingle is used to tokenize the whole file that is used for the encryption process. For instance, the group keys in Table 1, k1 is applied to block token 1 and k2 is applied for the block of token 2 and k3 is applied to block token 3 and so on.

Each encryption process needs to encryption key and plain text to be encrypted. In this paper, the keys generated based on Minhash function and the plain text is the data after applying the k-shingle. The group keys with the data is the first step of the process of merging the encryption algorithms on the Blowfish algorithm and then the output of them represents the input of the AES algorithm with the same keys used for each shingle. Figure 1 illustrates the main steps of the hybrid cryptosystem.

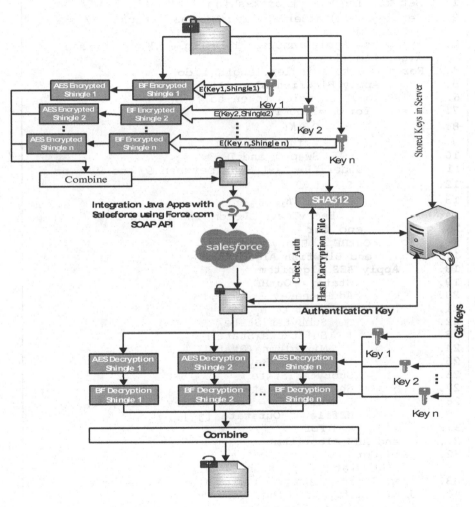

**Fig. 1.** The main steps of the hybrid cryptosystem

Algorithm (2) shows the main pseudo code for these steps:

Algorithm (2):Hybrid Cryptosystem Algorithm Based Minhash Generated Key

Input:  D (Input Data as a plaintext), $Ki = \{K1,....,Kn\}: K \subset$
$h(x_i) = (ax_i + b \bmod p)$ (Group of Keys from Minhash)
Output: HEfile(Hybrid Encryption file) , OutBF (output of

Blowfish) ,Outstate (output state of AES)
Begin
1.   Set k value   // k of k-shingle
2.   Get keys (*Ki*) after applied Minhash steps   // see Algorithm (1)
3.   shingles← Apply function K-shingle (K,D)   //see Algorithm (1)
4.   **For** j ← 0 to shingles.length() **do**
5.         **Apply Blowfish Algorithm**
6.         Split shingles token to $D_R$ and $D_L$
7.          **for i** =1 to **16do**
8.              $D_L \leftarrow D_L \oplus Ki$
9.                $D_R \leftarrow$ fun [$D_L$] $\oplus$ $D_R$
10.               Swap $D_R$ and $D_L$
11.            Under the last Swap between $D_R$ and $D_L$
12.            $D_R \leftarrow D_R \oplus K_{17}$
13.              $D_L \leftarrow D_L \oplus K_{18}$
14.            EncryToken$_i$←Recombine $D_L$ and $D_R$
15.         **end for**
16.         OutBF$_j$ ← EncryToken$_i$ ;
17.         **end Blowfish Algorithm**
18.      **Apply AES Algorithm**
19.         State$_j$ ← OutBF$_j$
20.         AddRoundKey(Stat $_j$ , $K_0$)
21.         **for** i← 1 to n$_{r-1}$**do**
22.             SubBytes(State$_i$)
23.             ShiftRows(State$_i$)
24.             MixColumns(State$_i$)
25.             AddRoundKey(State$_i$)
26.         SubBytes(State$_i$)
27.         ShiftRows(State$_i$)
28.         OutState[j] ← AddRoundKey(State$_j$, K$_j$)
29.         **HEfile ← Outstate [j ] ;**
30.         **end For**
31.    **end AES Algorithm**
32.     **End For**
     /* Integration app with Cloud
33.   Open Secure Connection with Cloud Salesforce
34.   Upload HEfile to Cloud Salesforce
35.  // Key exchange
36.   Hash value ← SHA512( **HEfile**)
37.Stored Hash value with *Ki* in index table database
**38.  End Algorithm**

## 5.3   The Key Management Phase

The main important step in the proposed cryptosystem is the key management. The distribution of the symmetric keys store the keys in a third part trusted server as the database to exchanges. This server is distributed in branches of the organization. A Hash function using SHA512 is implemented to generate Hash codes from the encrypted files which are going to be used for accessing the stored keys in decryption phase. The SHA512 Hash output is a message digest of 512 bits length. It is stored with the group keys generated from the key generation in Sect. 5.1 in the trusted server and it is used for authentication from the authorized client that needs to get keys. Figure 2 shows the proposed method of the key distribution in this paper.

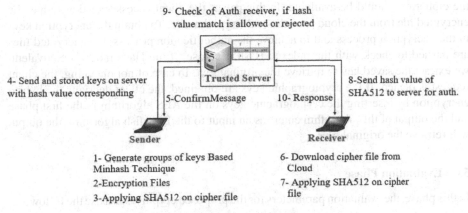

9- Check of Auth receiver, if hash value match is allowed or rejected

4- Send and stored keys on server with hash value corresponding

**Trusted Server**

5-ConfirmMessage        10- Response

8-Send hash value of SHA512 to server for auth.

**Sender**

**Receiver**

1- Generate groups of keys Based Minhash Technique

2-Encryption Files

3-Applying SHA512 on cipher file

6- Download cipher file from Cloud

7- Applying SHA512 on cipher file

**Fig. 2.**  The steps of key distribution

## 5.4   Integrating the Proposed Cryptosystem with the Cloud Phase

Integrating the proposed cryptosystem implemented in Java with the cloud Salesforce using the Force.com SOAP API. This work is implemented using the API that integrates the Java programing language with the Force.com Simple Object Access Protocol (SOAP) Applications Programming Interface (API) using the enterprise Web Service Definition Language (WSDL) for the client access and implementation. The TLS protocol is used to establish a secure connection to the cloud. TLS protocol provides a reliable end to end secure service. The following code illustrates adding TLS for the secure connection between the client application and the Salesforce cloud.

$$System.\,setProperty("https.\,protocols", "TLSv1, TLSv1.1, TLSv1.2");$$

This configuration is presented to create connection with the cloud implementation using Java program. It is used in several libraries in Salesforce to open connection through web service which is shown in code as follow:-

```
import com.sforce.soap.enterprise.Connector;
```

import com.sforce.soap.enterprise.EnterpriseConnection;

import com.sforce.soap.enterprise.SaveResult;

import com.sforce.soap.enterprise.sobject.Attachment;

import com.sforce.soap.enterprise.sobject.SObject;

import com.sforce.ws.ConnectionException;

import com.sforce.ws.ConnectorConfig;

## 5.5 The Hybrid Decryption Phase

To decrypt the ciphertext in any symmetric encryption system, the encryption key and the ciphertext should be available. In this paper, the client accesses or downloads the encrypted file from the cloud provider to decrypt that file. To obtain the encryption keys in the decryption process and to achieve the authentication process, the encrypted files are hashed to check with the indexed Hash database. If the Hash index is equivalent, we extract the saved key to retrieve the original text. In case of not matching, the client request is denied. If the cryptographic keys are obtained, the client decrypts the hybrid encryption by inserting the cryptographic keys on the AES algorithm in the first phase and the output of this algorithm enters as an input to the Blowfish algorithm, the output will retrieve the original files.

## 5.6 Evaluation Phase

In this phase, the evaluation parameters for the proposed cryptosystem use the following:

**Encryption Time:** It refers to the amount of time to convert data from plaintext to cipher text depending on the key size and data block size. In this work, the encryption/decryption time is calculated through Eq. (2).

$$Executiontime = EndTime - StartTime \qquad (2)$$

**Throughput:** It defines the number of bits which can be encrypted or decrypted during one unit of time. The Throughput is calculated by using Eq. (3) to calculate the speed of algorithms [27].

$$Throughput = filesizeinbyte/EncryptionTime \ (sec) \qquad (3)$$

**Memory Used:** The memory used is the space that is reserved for algorithms' implementation. It mainly depends on the number of operations in algorithm, initialization vectors and the type of mode operations [27].

**Avalanche Effect:** Avalanche effect means any change in plaint text or key leads to change in cipher text. It can be calculated using Eq. (4) by changing bits in plain text with constant key [27].

$$Avalanche effect = \frac{no. \ of \ flippedbitsincipheredtext}{no. \ of \ bitsincipheredtext} * 100 \qquad (4)$$

**Entropy:** It is calculated using the Shanon law. Equation (5) is used to calculate the entropy. It is a measure of randomness which indicates the power of the cipher algorithms against attack [28].

$$H(x) = -\sum_{i=0}^{n-1} p(x_i) \log(p(x_i)) \tag{5}$$

where H(x) means entropy for encryption shingles, xi is cipher text (single token). p is the probability of an occurrence of xi of shingling. The normalized entropy value (0.5) refers to the ideal value between the values of zero which indicates similarity and a value of 1 which indicates randomness.

Table 2 shows the results of the implementation of the hybrid cryptosystem for evaluations metrics from Eqs. (1–5).

**Table 2.** Evaluation parameters of the hybrid cryptosystem

| File size (bytes) | Hybrid encryption | | | |
|---|---|---|---|---|
| | Time (sec) | Throughput (Byte/sec) | Memory usage (KB) | Entropy (Bit) |
| 25885 | 1.254 | 2063.795853 | 107264 | 6.285402219 |
| 51697 | 2.29 | 22575.10917 | 27321 | 7.672425342 |
| 157214 | 7.107 | 22121.00746 | 62950 | 9.878050913 |
| 364827 | 16.412 | 22229.28345 | 78509 | 11.33817925 |
| 513983 | 24.243 | 21201.29522 | 109407 | 11.65418847 |
| 637950 | 28.882 | 22088.15179 | 162966 | 11.07815081 |
| 1278548 | 60.169 | 21249.28119 | 638561 | 12.42337858 |
| 2240213 | 103.712 | 21600.3259 | 908512 | 13.51286389 |
| 3600365 | 157.786 | 22818.02568 | 505812 | 13.40274562 |

Table 3 shows the evaluation parameters of hybrid cryptosystem decryption.

In addition to the results mentioned above for the algorithms, the performance evaluation using avalanche effect is considered a good security indicator for these algorithms. Table 4 shows the results for the avalanche effect.

It is clearly shown from this Table 4 that the hybrid cryptosystem (BF-AES) has the highest values and it fulfills the best results in comparison with other algorithms. AES and Blowfish have good values of the avalanche effect which make these algorithms more robust. Avalanche effect values are calculated by using Eq. (4) for several files with content and length different and find the average of values.

The obtained results pave the way to the implementation of a hybrid cryptosystem based on two block cipher Algorithms (Blowfish and AES).

**Table 3.** Evaluation parameters of the hybrid cryptosystem decryption

| File size (bytes) | Hybrid cryptosystem decryption | |
| --- | --- | --- |
| | Time(Sec) | Throughput (Byte/Sec) |
| 25885 | 0.254 | 10188.97638 |
| 51697 | 2 | 25848.5 |
| 157214 | 6.307 | 24926.90661 |
| 364827 | 13.562 | 26900.67837 |
| 513983 | 19.643 | 26166.21697 |
| 637950 | 25.282 | 25233.36761 |
| 1278548 | 43.969 | 29078.39614 |
| 2240213 | 91.802 | 24402.66007 |
| 3600365 | 124.896 | 28826.90398 |

**Table 4.** Average of Avalanche effect in changing bits in plain text with constant keys

| Method | No. bits filliped | Avalanche effect (%) |
| --- | --- | --- |
| AES | 70 | 96.30682 |
| Blowfish | 195 | 95.46543 |
| BF-AES | 291 | 98.45602 |

## 6 Conclusion

This paper provided an efficient way of using the principles of the Minhash method to implement and generate a set of keys, which act as input keys to a hybrid encryption approach for a large number of documents or files that are encrypted and uploaded to cloud computing. The main conclusion of this work is that the robust keys generation method is based on the principles of K-shingle and Minhash technique. The obtained results show that the Blowfish and AES are more efficient and suitable for this work. It is efficient to conduct speed, confidentiality and integrity with high **priority**. The entropy performance criteria satisfied a high score of randomizations. The hybrid encryption algorithms are implemented to encrypt files by the combination of AES and Blowfish algorithms. This combination shows a better performance, in terms of security-based entropy and avalanche effect, than that of each of AES and Blowfish when used separately. In addition, a robust authentication key distribution uses the hash code and database server.Java App is integrated with the Salesforce cloud provider using SOAP protocol through WSDL and forc.com web service connector API. The free cloud computing feature is used to implement this work by integrating the cloud computing libraries with Java programming. Finally, decryption work is integrated by requesting keys previously stored in the database on the server.

# References

1. Andress, J.: The basics of information security: understanding the fundamentals of InfoSec in theory and practice. Syngress (2014)
2. Blumenthal, M.: Encryption: strengths and weaknesses of public-key cryptography. dalam Computer Science Research Symposium, Villanova, pp. 1–7 (2007)
3. Stallings, W.: Cryptography and Network Security: Principles and Practice, 6th edn. (2014). ISBN 9780133354690, 0133354695
4. Mahajan, P., Sachdeva, A.: A study of encryption algorithms AES, DES and RSA for security. Glob. J. Comput. Sci. Technol. Netw. Web Secur. **13**(15), 33–38 (2013)
5. Mell, P., Grance, T.: The NIST definition of cloud computing (draft) (2011)
6. Jayapandian, N., Zubair Rahman, A.M.J.Md., Radhikadevi, S., Koushikaa, M.: Enhanced cloud security framework to confirm data security on asymmetric and symmetric key encryption. In: 2016 World Conference on Futuristic Trends in Research and Innovation for Social Welfare (Startup Conclave), pp. 1–4. IEEE (2016)
7. Tweneboah-Koduah, S., Endicott-Popovsky, B., Tsetse, A.: Barriers to government cloud adoption. Int. J. Manag. Inf. Technol. (IJMIT) **6**(3), 1–16 (2014)
8. Abutaha, M.S., Amro, A.A.: Using AES, RSA, SHA1 for securing cloud. In: International Conference on Communication, Internet and Information Technology, Madrid, Spain (2014)
9. Liu, W.: Research on cloud computing security problem and strategy. In: 2012 2nd International Conference on Consumer Electronics, Communications and Networks (CECNet), pp. 1216–1219. IEEE (2012)
10. Benioff, M.: Salesforce, "Salesforce.com" (2001). https://www.salesforce.com/
11. Mahalle, V.S., Shahade, A.K.: Enhancing the data security in cloud by implementing hybrid (RSA & AES) encryption algorithm. In: 2014 International Conference on Power, Automation and Communication (INPAC), pp. 146–149. IEEE (2014)
12. Liang, C., Ye, N., Malekian, R., Wang, R.: The hybrid encryption algorithm of lightweight data in cloud storage. In: 2016 2nd International Symposium on Agent, Multi-Agent Systems and Robotics (ISAMSR), pp. 160–166. IEEE (2016)
13. Chauhan, A., Gupta, J.: A novel technique of cloud security based on hybrid encryption by blowfish and MD5. In: 2017 4th International Conference on Signal Processing, Computing and Control (ISPCC), pp. 349–355. IEEE (2017)
14. Prathana Timothy, D., Kumar Santra, A.: A hybrid cryptography algorithm for cloud computing security. In: 2017 International Conference on Microelectronic Devices, Circuits and Systems (ICMDCS), pp. 1–5. IEEE (2017)
15. Mahmud, A.H., Angga, B.W., Tommy, Marwan, A.E., Siregar, R.: Performance analysis of AES-blowfish hybrid algorithm for security of patient medical record data. J. Phys. Conf. Ser. **1007**(1), 1–4 (2018)
16. Salama Abdelminaam, D.: Improving the security of cloud computing by building new hybrid cryptography algorithms. Int. J. Electron. Inf. Eng. (IJEIE) **8**(1), 40–48 (2018)
17. Sangeetha, S.K.B., Vanithadevi, V., Rathika, S.B.: Enhancing cloud security through efficient fragment based encryption. Int. J. Pure Appl. Math. **118**(18), 2425–2436 (2018)
18. Lee, B., Dewi, E.K., Wajdi, M.F.: Data security in cloud computing using AES under HEROKU cloud. In: The 27th Wireless and Optical Communications Conference (WOCC2018) Data, pp. 1–5. IEEE (2018)
19. Kartit, Z., et al.: Applying encryption algorithm for data security in cloud storage. In: Sabir, E., Medromi, H., Sadik, M. (eds.) Advances in Ubiquitous Networking. LNEE, vol. 366, pp. 141–154. Springer, Singapore (2016). https://doi.org/10.1007/978-981-287-990-5_12
20. Ismail, M., Yusuf, B.: Ensuring data storage security in cloud computing with advanced encryption standard (AES) and authentication scheme (AS). Int. J. Inf. Syst. Eng. (IJISE) **4**(1), 18–39 (2016)

21. Ebady Manna, M., Ali Mohammed A, M.: Data encryption scheme for large data scale in cloud computing. J. Telecommun. Electron. Comput. Eng. **9**(2–12), 1–5 (2017)
22. Forouzan, B.A.: Cryptography and Network Security, 1st edn. McGraw-Hill Publishing Company, New York (2007). ISBN 13-978-0-07-066046-5, 10-0-07-066046-8
23. Paar, C., Pelzl, J.: Understanding Cryptography: A Textbook for Students and Practitioners. Springer, Heidelberg (2009). https://doi.org/10.1007/978-3-642-04101-3
24. Das, A., Datar, M., Garg, A., Rajaram, S.: Google news personalization: scalable online collaborative filtering. In: Proceedings of the 16th International Conference on World Wide Web, pp. 271–280. ACM (2007)
25. Leskovec, J., Rajaraman, A., Ullman, J.D.: Mining of Massive Datasets, 2nd edn. Cambridge University Press, Cambridge (2014)
26. Reddy Nagireddy, S.: Scalable techniques for similarity search (Master's Project). San Jose State University SJSU ScholarWorks (2015)
27. Patidar, G., Agrawal, N., Tarmakar, S.: A block based encryption model to improve avalanche effect for data security. Int. J. Sci. Res. Publ. **3**(1), 1–4 (2013)
28. Cachin, C.: Entropy measures and unconditional security in cryptography. Swiss Federal Institute Of Technology Zurich (Doctoral dissertation, ETH Zurich (1997)

# Machine Learning Approaches

# Digitally Modulated Signal Recognition Based on Feature Extraction Optimization and Random Forest Classifier

Batool Abd Alhadi[1(✉)], Taha Mohammed Hasan[1(✉)], and Hadi Athab Hamed[2(✉)]

[1] College of Sciences, University of Diyala, Diyala, Iraq
zainabkadam10@gmail.com, Dr.tahamh@sciences.uodiyala.edu.iq
[2] Technical College-AL Mussaib, Al-Furat Al-Awsat Technical University, Babil, Iraq
Com.had@atu.edu.iq

**Abstract.** In the past decades, there was a growing need for the automatic classification that is related to digital signal formats, that also appears to be on going tendency in the future. Automatic modulation recognition (AMR) is considered to be of high importance in military and civil applications and communication systems. The recognition regarding the received signal modulation can be defined as a transitional stage between detection and demodulation of signals. in this paper, several features which are associated with the received signal will be extracted and used. Which is of high importance in increasing the AMR's effectiveness. Algorithms from Chicken Swarm optimization and Bat Swarm optimization were used to improve the features of modulated signals and thus increase the accuracy of the classification. It then classifies the features of the modified signals resulting from the optimization algorithms by a random forest. The results showed that swarm Chickens algorithm performs better than the Bat swarm algorithm, even at level SNR low, Results From chicken Algorithm & Random forest classifier were Accuracy of classification 95% while Accuracy of classification From Bat Algorithm & Random forest 91%.

**Keywords:** Chicken swarm optimization algorithm · Random forests classifier · Bat swarm optimization algorithm

## 1  Introduction

AMR can be defined as a method which report the modulation types regarding the received signals in automatic way. Modulation recognition is very essential in communication intelligence (COMINT). In COMINT one needs to recognize the modulation scheme to demodulate the signal information content. Furthermore, classifying right modulation type is helpful to choose an appropriate type of jamming to transmit. Signal modulation recognition is an interdisciplinary composed of signal extraction, signal processing, signal analysis, pattern recognition and so on. As the complex nature of the received signals, there are still many problems to be solved urgently-feature extraction and modulation classifier [1].

© Springer Nature Switzerland AG 2020
A. M. Al-Bakry et al. (Eds.): NTICT 2020, CCIS 1183, pp. 75–84, 2020.
https://doi.org/10.1007/978-3-030-55340-1_6

In [2]. In this study, the researchers selected suitable features for the modified input signal and used the neural network algorithm. The presence of Gaussian noise $-5$ dB to 20 dB. The results showed that the use of the NN algorithm resulted in a significant increase in the precision of recognition of the type of modification.

In [3] using modified artificial bee colony (MABC) with Enhanced Particle Swarm Optimization (EPSO) for predicting the infiltration detection. The algorithms will be combined for the purpose of achieving improved results of optimization and obtain accuracy of classification.

In [4] use bilateral support vector machine (SVM) for the purpose of determining digital modulation approaches. Modulation types consist of: BASK, BFSK, BPSK, 4ASK, 4FSK, QPSK and 16-QAM. Simulation demonstrates the higher abilities regarding suggested features in the digitally modified signal classification Even at a low SNR level.

In [5] Artificial Neural Network (ANN) algorithm was utilized for classifying the modification. In the presented study, the system of AMC have the ability of distinguishing between USB, QPSK, LSB, 8 PSK, and AM modulation. The rate of accuracy that is related to the presented system in modulation classification process without the use of nonlinear conversions is 65.5% on the signal quality of 10 dB. Therefore, the precision of AMC with the use of nonlinear transformations on incoming signal is about 88.8% on signal quality 10 dB.

In [6] using the decision tree a dual-carrier bus system support has been trained on the features which have been extracted from high-order stacking tool. The results have indicated that the average accuracy of identifying formatting signals can be realized more than 94% when SNR is $-5$ dB.

In fact, paper submitted by Almaspour and Moniri [2] under title "Automatic modulation recognition and classification for digital modulated signals based on ANN algorithms". The closest research with our study, we have used Same types of modified signals test the performance of the Classification. We used (CSO, BA) algorithm and Random Forest Classifier To improve system performance. While the researchers selected suitable features for the modified input signal and used the neural network algorithm to classified the signal, better results were obtained.

The major aim of this study is optimizing the features of the modulated signals, thus reducing the signal characteristics and thereby increasing the accuracy of the system in detection and identification of the signal type, using two types of Swarm optimization algorithms and comparing their performance to detect the best in strengthening the features. In this work, Random forest is as classified. Also, one of the main issues is selecting the suitable feature set. In previous paper, usually are numerous Features used to classify the composition leading to Improved efficiency. So many features are not effective enough one of the main reasons leads to restrictions Most of the techniques recognize a digital signal The features are related [7]. The general structure of this study is as follows. After introduction, features extraction and algorithms of optimization are going to be reviewed in Sect. 2, the classified will be presented in Sect. 3. Section 4 Provide some simulation results, and lastly the conclusions of the study will be presented in Sect. 5.

## 2 Feature Extraction

The initial stage in extracting features is to model the modified digital signals for extracting the signal features. For the purpose of analyzing the implementation regarding the metering system [2] Various digital signal types have a lot of features. So find appropriate features to recognize the scheme of digitally modulated signals are a serious problem. The main features of the configuration classification should be identified in the pattern recognition approach. These features must be Strong properties are sensitive to modulation types and are not sensitive with SNR variance. Instantaneous. Features such as Moments and cumulants were calculated. These were used to classify signals Features [8]. $k^{th}$-order cumulant can be written in the following way based on the HOCs property

$$C_{k,n(x(t))} = C_{k,n(s(t))} + C_{k,n(n(t))} \tag{1}$$

As the white Gaussian noise's value is considered to be larger compared to $2^{nd}$ order cumulant constant to 0, i.e.

$$C_{k,n(n(t))} = 0, K > 2 \tag{2}$$

Thus, the non-Gaussian signal that has been received and consist of white Gauss noise is going to be converted to cumulant processing, and thus the impact of noise will be removed. Typically, HOC will be defined as function related to the high-order moments (HOMs) of the signal. For complex $k^{th}$-order stationary stochastic procedure S(t) with zero average value, which its $p^{th}$-order mixing moment [9] is

$$M_{pq=E[S\,(t)^{p-q}S^*(t)^q]} \tag{3}$$

In which p will be referred to as the moment order and [.]* represents complex conjugation. Symbolism for $k^{th}$-order of cumulant is comparable to that of pth-order moment. In particular:

$$C_{pq} = Cum\left[\underbrace{s(t),\ldots,s(t)}_{p-q},\underbrace{S^*(t),\ldots,S^*(t),}_{q}\right] \tag{4}$$

Thus, depending on Eq. (3) and Eq. (4), one might define an association between HOMs and HOCs of $2^{nd}$ to $8^{th}$ order cumulants are indicated in (Table 1).

### 2.1 Chicken Swarm Algorithm

Chicken Swarm Optimization (CSO) can be defined as a novel intelligent bionic algorithm suggested based on different behaviors of chicks, hens, and cocks in their search for food. In CSO the chicken swarm in the search space will be mapped as particular particle individual. Chicken particle swarm, hen particle swarm, and cock particle swarm undergo sorting depending on the particle's fitness value, different search mode is utilized by each sub swarm [10] In CSO a lot of particles with optimum fitness will be chosen as the cock particle swarm, that is specified via

$$X_{i,j}^{t+1} = X_{i,j}^t + randn\left(0, \sigma^2\right).X_{i,j}^t \tag{5}$$

**Table 1.** The correlation between the HOCs and HOMs.

| HOCS | HOMS expresion |
|------|----------------|
| Second order cumulant | $C_{2,0}$  $M_{2,0}$ |
|  | $C_{2,1}$  $M_{21}$ |
| Fourth order cumulants | $C_{4,0}$  $M_{4,0} - 3M_{2,0}^2$ |
|  | $C_{4,1}$  $M_{4,1} - 3M_{4,0}$, $M_{21}M_{4,1} - |M_{21}|^2 - 2M_{2,1}^2$ |
| Sixth order cumulant | $C_{6,0}$  $M_{6,1} - 15M_{4,0}M_{2,0}M_{4,1} + 30M_{2,1}^3$ |
|  | $C_{6,3}$  $M_{6,3} -$ |
|  | $M_{4,1}M_{2,1} + 12M_{2,1}^3 - 3M_{2,0}M_{4,3} - 3M_{2,2}M_{4,1}\ 18M_{2,0}\ M_{1,2}M_{2,2}$ |
| Eight order cumulant | $C_{8,0}$  $M_{6,3} - 28M_{6,0}M_{2,0} - 35M_{4,0}^2 + 420M_{4,0}M_{2,0}^2$ |
|  | $-630M_{2,0}^4$ |

In which $X_{i,j}^{t+1}$ and $X_{i,j}^t$ representing the location of the $j^{th}$ dimension of the $i$-th particle in $t + 1$ and $t$ iterations respectively and randn $(0, \sigma^2)$ can be defined as arbitrary number of the Gauss distribution whose variance $= \sigma^2$. The parameter $\sigma^2$ might be estimated through

$$\sigma^2 = \begin{cases} 1, & fit_i < fit_k \\ \exp\left(\frac{(fit_k - fit_i)}{(|fit_i| + \xi)}\right), & fit_i \geq fit_k \end{cases} \tag{6}$$

In which $i, k \in [1, r$ size$]$ and $i \neq k$. $r$ size are representing the number of the cock swarms. 〚fit〛 _i and 〚fit〛 _k are representing the fitness values regarding the cock particle $i$ and $k$ respectively; $\xi$ representing the number that is small enough. Furthermore, the majority of particles that have optimum fitness have been chosen as the hen swarms. Its random search is achieved through cocks of hen population and that of others, that could be specified as

$$X_{i,j}^{t+1} = X_{i,j}^t + s2rand.\left(X_{r1,j}^t, X_{i,j}^t\right) + .s2\,rand.\left(X_{r2,j}^t, X_{i,j}^t\right) \tag{7}$$

In which $X_{(r1,j)}^t$ and $X_{(r1,j)}^t$ are representing the position regarding the cock individual $r1$ in hen population $xi$ and cock individual $r2$ in other population, respectively. rand is uniform random number over $[0, 1]$. $S1$ and $S2$ refer to the weight estimated via

$$s1 = exp\left(\frac{(fit_{i-fit_{r1}})}{(|fit_i| + \xi)}\right)$$

$$S2 = exp(fit_{r2} - fit_i) \tag{8}$$

In which $fit_i$ and $fit_k$ are, respectively, cock's individual fitness value $r1$ in population of the hen $xi$ and cock individual $r2$ in other population. Every individual, apart from hen swarm and cock swarm, are specified as chick swarm. Its mode of search mode follow that of the hen swarm, that is specified via

$$X_{i,j}^{t+1} = X_{i,j}^t + FL.\left(X_{m,j}^t - X_{i,j}^t\right), FL \in [0, 2] \tag{9}$$

## 2.2 Bat Swarm Algorithm

Bat Algorithm (BA) can be defined as swarm intelligence based algorithm, it depends on the echolocation behavior regarding the micro bats. When the bats are flying and hunting, they will be emitting certain ultrasonic, short, pulses to environment and list to their echoes. The studies have indicated that the information they obtain from the echoes are going to help the bat to construct accurate image regarding the environment around them and determining the shape, distance, and the location of the prey. This echolocation behavior of the bats is very important in finding the prey as well as distinguishing a lot of insects' types in total darkness. Previous studies indicated that the Bat Algorithm have the ability of solving un-constrained and constrained optimization problems with a lot more robustness and effectiveness in comparison to Genetic Algorithm and Particle Swarm Optimization [11]. The following describes the execution steps of the typical BA:

**Step 1:** For each bat, initialize the position, velocity, and parameters and randomly generate the frequency.
**Step 2:** Update the position and velocity of each bat.
**Step 3:** For each bat, generate a random number ($0 < rand_1 < 1$). Update the temp position and calculate the fitness value for corresponding bat if $rand_1 < r_i(t)$
**Step 4:** For each bat, generate a random number ($0 < rand_2 < 1$),, Update $A_i(t)$ and $r_i$ ($t$) if $rand_2 < A_i(t)$ and $f(X_i(t)) < f(p(t))$.
**Step 5:** Sort each individual based on fitness values and save the best position.
**Step 6:** The algorithm is finished if the condition is met, otherwise, move on to Step 2.

# 3 Classifier

## 3.1 Random Forest

Random forest can be considered as ensemble classifier which contain a lot of decision trees and output the class which is mode of class's output via separate trees. The term was derived from the random decision forests which has been initially suggested via [12]. The approach combine feature's random selection and the Breiman's "bagging" approach, presented. For the purpose of constructing set of decision trees with controlled variation. The random forests combine the tree predictors in a way that every one of the trees depend on the random vector values that are sampled independently and with same distribution for every forest tree. The generalization error regarding forest of the tree classifiers depend on strength regarding distinct trees in forest and association between them.

Random forests operate effectively on huge databases that can handle huge volumes of input variables with no deletion. It offers assessment regarding the significant variables in classification. Random forest in considered to be un-biased toward the assessment regarding the generalized error throughout forest formation. The algorithm also effectively calculate the missing data as well as preserving precision with approaches used to balance the errors in un-balanced class population datasets. The Resultant forests could be managed as inputs to future datasets. It provides information regarding the association between classification and variables. It operates extremely well in outlier detection, labeling of un-supervised clustering and data views Streaming Random Forest learning Algorithm. Random forest algorithm consist of the following steps:

**Step 1:** Assuming that S is the number of training samples, whereas P is the number of variables in the classifier.
**Step 2:** Assume that p is the number of input variables that are utilized for determining decision at a node of the tree where p must be considerably smaller than P.
**Step 3:** Choose a training set for certain tree via choosing S times with replacement from every S training sample available. Via predicting the classes, the remaining samples are utilized for estimating the error of the tree.
**Step 4:** In order to make a decision at one of the nodes, randomly choose p variables for every one of the tree nodes. Calculate the optimal split in the training dataset according to the p variables.
**Step 5:** Every one of the trees will be grown to its maximum possible level so that there isn't any more pruning [13]. Figure 1 shows the structure of Random Forest.

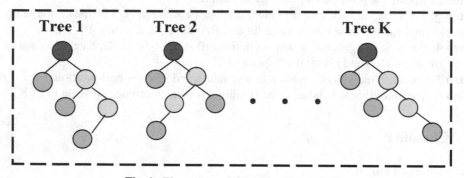

**Fig. 1.** The structure of Random Forest [14]

## 4 Proposed System

Was used Nine types (QAM8, QAM16, QAM32, QAM 64, QAM 128, QAM 256, PSK2, PSK4, ASK2) of Modulated signals Which we have generated in the MATLAB Program within a level of SNR ranging from $(-2, -1, 0, 1, 2, 3, 4, 5)$ dB. After extracting the generated signal features. The extracted features were collected in a data matrix to improved it. This data Go through two stages: first phase is starting with optimization signal using two algorithms (CSO, BA). The second phase: is that takes the output from phase 1 as an input to random forest the used as a classifier to determine the accuracy of the classification and prediction of the type of signal. Figure 2 shows the proposed system. Where the algorithms used in the proposed system are mentioned in Sect. 3.

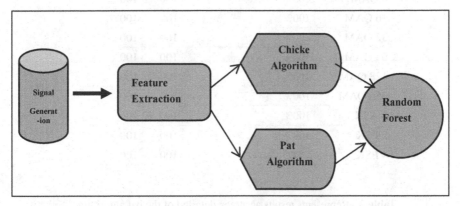

**Fig. 2.** The schematic of the proposed approach

## 5 Experimental Result

In this section, the results that have been obtained after Optimization the signal features modulated by the algorithm (Chicken & Random forest) and the results obtained after optimizing the features using a Java (Bat & Random forest) algorithms. The results of both techniques were compared.

1. The results obtained after execution of the SNR $(-2, -1, 0, 1, 2, 3, 4, 5)$ dB in Random Forest classification algorithm and Chicken algorithm are listed in the Table 2 (Fig. 3).
2. The results obtained after execution of the SNR $(-2, -1, 0, 1, 2, 3, 4, 5)$ dB in Random Forest classification algorithm and Chicken algorithm are listed in the Table 3.

**3. Comparison Between Performance (Chicken, Bat)**
The results drawn from Two different techniques (chicken, Bat) optimization algorithm) and classification by Random Forest. Show chicken algorithm more efficiently than Bat algorithm has higher performance and higher classification accuracy. Been focusing on three basic criteria (precision, Recall, F-Measure) (Fig. 4).

**Table 2.** Represents results accuracy detailed of the chicken algorithm

| Type of signal | Classification criteria | | |
|---|---|---|---|
| | Accuracy of classification | Recall | F-measure |
| 8 QAM | 100% | 100 | 100 |
| 16 QAM | 100% | 100 | 100 |
| 32 QAM | 100% | 100 | 100 |
| 64 QAM | 100% | 100 | 100 |
| 128 QAM | 100% | 100 | 100 |
| 256 QAM | 100% | 100 | 100 |
| 2 ASK | 100% | 100 | 100 |
| 2 PSK | 100% | 100 | 100 |
| 4 PSK | 54% | 100 | 100 |

**Table 3.** Represents results accuracy detailed of the bat algorithm

| Type of signal | Classification criteria | | |
|---|---|---|---|
| | Accuracy of classification | Recall | F measure |
| 8 QAM | 100% | 100 | 100 |
| 16 QAM | 100% | 100 | 100 |
| 32 QAM | 100% | 100 | 100 |
| 64 QAM | 100% | 100 | 100 |
| 128 QAM | 57% | 50 | 63 |
| 256 QAM | 100% | 100 | 100 |
| 2 ASK | 100% | 100 | 100 |
| 4 PSK | 66% | 75 | 70 |
| 2 PSK | 100% | 87 | 93 |

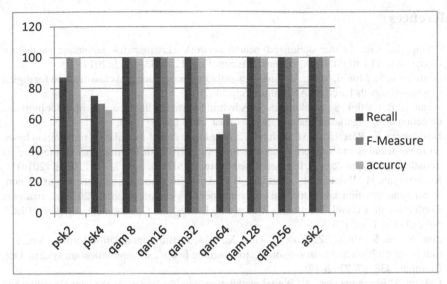

**Fig. 3.** Represent the result obtain after optimization feature by Bat algorithm

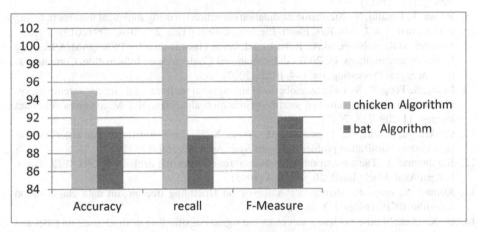

**Fig. 4.** Comparison between (Bat & CSO)

## 6 Conclusion

In this paper, Nine types of electromagnetic signals embedded in the MATLAB program were created within an SNR level ranging from (−2, −1, 0, 1, 2, 3, 4, 5) dB. Then we extract the statistical characteristics (moment, cumulant) of the signals, above. We improve the features, by two way Chicken swarm optimization algorithm and Bat Swarm Optimization algorithm. Using the Random Forest as a classifier for both. The results were compared of the proposed methods, the chicken swarm algorithm outperforms Bat Swarm algorithm by predicting the signal type and obtaining the highest classification accuracy even at a low SNR level of about 95%.

# References

1. Cheng, L., Liu, J.: An optimized neural network classifier for automatic modulator recognition. TELKOMNIKA Indones. J. Electr. Eng. **12**, 1343–1352 (2014)
2. Almaspour, S., Moniri, M.R.: Automatic modulation recognition and classification for digital modulated signals based on ANN algorithms. **3** (2016)
3. Amudha, P., Karthik, S., Sivakumari, S.: A hybrid swarm intelligence algorithm for intrusion detection using significant features. Sci. World J. **2015** (2015). 15 p.
4. Hassanpour, S., Pezeshk, A.M., Behnia, F.: Automatic digital modulation recognition based on novel features and support vector machine. In: 2016 12th International Conference on Signal-Image Technology & Internet-Based Systems (SITIS), pp. 172–177. IEEE (2016)
5. Kurniansyah, H., Wijanto, H., Suratman, F.Y.: Automatic modulation detection using non-linear transformation data extraction and neural network classification. In: 2018 International Conference on Control, Electronics, Renewable Energy and Communications (ICCEREC), pp. 213–216. IEEE (2018)
6. Sun, X., Su, S., Huang, Z., Zuo, Z., Guo, X., Wei, J.: Blind modulation format identification using decision tree twin support vector machine in optical communication system. Opt. Commun. **438**, 67–77 (2019)
7. Hakimi, S., Ebrahimzadeh, A.: Digital modulation classification using the bees algorithm and probabilistic neural network based on higher order statistics. Int. J. Inf. Commun. Technol. Res. **7**, 1–15 (2015)
8. Bagga, J., Tripathi, N.: Automatic modulation classification using statistical features in fading environment. Int. J. Adv. Res. Electr. Electron. Instrum. Eng. **2**, 3701–3709 (2013)
9. Mirarab, M.R., Sobhani, M.A.: Robust modulation classification for PSK/QAM/ASK using higher-order cumulants. In: 2007 6th International Conference on Information, Communications & Signal Processing, pp. 1–4. IEEE (2007)
10. Liang, S., Feng, T., Sun, G.: Sidelobe-level suppression for linear and circular antenna arrays via the cuckoo search–chicken swarm optimisation algorithm. IET Microwaves Antennas Propag. **11**, 209–218 (2017)
11. Chakri, A., Khelif, R., Benouaret, M., Yang, X.-S.: New directional bat algorithm for continuous optimization problems. Expert Syst. Appl. **69**, 159–175 (2017)
12. Barandiaran, I.: The random subspace method for constructing decision forests. IEEE Trans. Pattern Anal. Mach. Intell. **20**, 832–844 (1998)
13. Kumar, A., Kaur, P., Sharma, P.: A survey on Hoeffding tree stream data classification algorithms. CPUH-Res. J. **1**, 28–32 (2015)
14. Li, K., et al.: Multi-label spacecraft electrical signal classification method based on DBN and random forest. PLoS ONE **12**, e0176614 (2017)

# Churners Prediction Using Relational Classifier Based on Mining the Social Network Connection Structure

Asia Mahdi Naser Alzubaidi[1]($\boxtimes$) (ID) and Eman Salih Al-Shamery[2]

[1] College of Science, University of Karbala, Karbala, Iraq
asia.m@uokerbala.edu.iq
[2] College of Information Technology, Babylon University, Babel, Iraq

**Abstract.** Customer churn prediction models aim to indicate the customers with the high tendency to churn, allowing for improved efficiency of customer retention operations and reduced costs associated with the attrite event. This paper proposed a data mining model to predict churn customers using Call Detail Records (CDR) data in the Telcom industry. CDR data are valuable for understanding the social connectivity between customers through call, message or chat graph but do not immediately provide the strength of their relations or present adequate information of how to churn node diffuse it is influenced to all neighbor nodes within the call graph. The main contribution of this paper is to propose a data mining model to predict Potential churners based on social ties strength and churn diffusion through network nodes. The model formulated by extracting the predefined number of social network communities that denoted the fundamental constructions for figuring out the connections structure of the call graph using Non-negative matrix factorization approach. Then, for each community quantifies the social ties strength using Node interaction and similarity algorithm. Then, incorporate social strength ties in an influence propagation model using are receiver-centric algorithm with Logistic Regression classifier to predict the set of the customers at risk of churn. Experiments conducted on Telecom dataset and adopting of AUC, ER, Accuracy, F-score, Lift, ROC, KS and H measure as evaluator metrics show that using of influence node diffusion based on ties strength boosted the performance of the proposed churn model between 84% to 91% in terms of the accuracy metric.

**Keywords:** Call graph · Churn prediction · Ties strength · Propagation model · Call Detail Records · Social network analysis · Churn diffusion · Community detection · Non-negative matrix factorization

## 1 Introduction

The Telecom industry is a highly technological sector which has developed tremendously over the past two decades as a result of the emergence and commercial success of both mobile Telecommunication and the internet [1]. In an era of mature markets and intensive competitive pressure, manage relationships with customers to increase the revenues are fundamental for companies and in business economics, this concept is

© Springer Nature Switzerland AG 2020
A. M. Al-Bakry et al. (Eds.): NTICT 2020, CCIS 1183, pp. 85–104, 2020.
https://doi.org/10.1007/978-3-030-55340-1_7

known as "Customer Relationship Management" (CRM), which is a business and profitable strategy that aims to ensure customer's satisfaction and loyalty. The companies which successfully apply CRM to their business it often improves their retention power and increases customer loyalty [2]. Telecom companies invest considerable resources to provide superior products and services to overcome their competitors, so many companies adopt new strategies to stay in competitive and deal with difficult problems to eliminate key threats from profits lost, inaccurate inter billing, fraud and churn events. Consequently, a powerful Business Intelligence solution e.g. analyzing and integrating in-depth data becoming effective strategies that help Telecom providers to reduce revenue losses [9]. A CDR data involves various details for every call, e.g. Caller, Callee, call/SMS duration, number of calls, number of SMS, etc. Based on these attributes, one can create a call graph with sender/receiver numbers as nodes and the number and duration of Calls/SMS as the weight of edges [1]. The weight of an edge captured the strength of the ties between a pair of nodes, the excessive-weight side indicates a robust tie, whilst the low-weight side is a weak one. As a result of the enormous computational complexity included in social relationships, the main hazard of social networking is the capacity to recognize patterns of social behavior between members and the influence of every event that occurs in terms of personal influence and the model derives from the view that the attributes of a user in a social networking are less primitive than their ties with different users within the same network [2]. Discovering the nature and strength of these relationship scan assist to understand the structures and dynamics of social graphics and provide an explanation for the real-world events from organizational structure to understanding the diffusion of churn process [3]. The recent researches on customer churn prediction argue that the churn influence can propagate or diffuse in the social network [4]. The propensity amount of some customers to churn has resulted from collected churn influences that can be obtained from their connections who've already churned. Hence, it has been proven that influence diffusion in call graph implemented in the phenomena of "word of mouth" and customer behaviors regarding churn event are studied with the incorporation of social networking context with the data mining method [5]. The Telecom network is complex enterprise, and the contents of customer communication must be preserved, consequently, it is hard to provide an explanation or quantify the propagation of churn influence among the circle customers [3]. Due to the complexity of developing measures such as call cost or even account profit, it is easier to simply use the cost of calls to outline the relationships between pairs in the network and it is important to note that call values vary according to the time of day and whether they occur on weekdays or weekends. Also, it is very common to find a big difference between two customer values in terms of the number of outgoing calls, incoming calls, or total call duration [6].

The structure of the article prepared in sections as illustrate in follows: Sect. 2, present the previous studies about customer churn prediction in the Telco sector. Methodology, model building, Data preprocessing, executed methods are described in Sect. 3. Section 4 illustrated the experimental implementation and outcomes of churn system are discussed. Conclusions considered in Sect. 5.

## 2  Related Works

Churners/non-churners classification model considered as predominant trouble for Telecom service providers and outlined as attrite customers because they go away to competitors. The capability of classifying the customer churn in advance, provide the Telcom company with appropriate insight to retain and maintains their customer. Variety of techniques have been applied to predict churn in Telecom companies with the incorporation of SNA that has been demonstrated of its importance by looking at the precise structuring and contenting of the ties-strength between customers and at the influence of network content, material and structure on model efficiency. See Table 1.

Phadke et al. [7] presented technique using SNA from the call graph to identify potential churners, They measure the strength of social ties among customers based on many attributes, and then apply the influence diffusion model on the call graph to decide the cumulative net influence of the churner node. This churn propagation model and other social factors were then combined with traditional metrics and machine learning methods are applied to calculate the propensity of churn for individual users. They concluded that the proposed approach performed well in terms of measurement of accuracy and was general and could be applied to any events that had an influence of diffusion.

Abd-Allah et al. [8] proposed a model for dyadic churn prediction using CDR data from probably the largest Telecommunications provider in Egypt. They built the SNA model based on founded the strong social relationships and the usage of these links for the diffuse of churn in a Telecommunications network. Experimental results exhibit that "call length" is probably the strongest purpose in promoting social ties, not like "call instances" that have little influence on individual connections. The outcome gained from the preliminary work consideration in churn modeling as a dyadic kind tested the importance of strong social links within the call graph.

Olle et al. [9] build a multi-branched network, a graph created by mobile users and a Linear Threshold application to model visitors deployed in the social network obtained to analyze the importance of social communication in spreading the influence of churn. Found that the spread of churn with the initial number of churners grows rapidly when there are communities in the network and are partly related to the structure of the communities through which the information is disseminated.

Pagare et al. [10] suggested utilizing social network attributes for customers with CDR data to predict churn users. The set of most influential nodes are expected by way of using the social network data, these churners could also be the one who may influence the others to discontinue utilizing a service provider. The results showed that the accuracy of churn prediction is developed and raised the quality of churn prediction procedure when combining the social network and get in touch with call log information of the customers for churn prediction making use of influence maximization.

Kamuhanda et al. [11] NMF algorithm was proposed to identify multiple local communities for single seeds selected randomly in communities, the number of communities is routinely calculated by the algorithm, they apply a breadth-first search to sample the input graph at multiple levels according to network connections intensity. Then, use the NMF in the adjacent matrix to estimate and obtain the communities in the network. The proposed system has been evaluated on actual-world networks and indicates excellent

**Table 1.** Summary of previous works

| No | Authors | Title | Journal | Year | Objective | Techniques | Dataset | Performance metrics |
|---|---|---|---|---|---|---|---|---|
| 1 | Phadke, Uzunalioglu, MendirattaKushnir, and Doran [7] | Prediction of Subscriber Churn Using Social Network Analysis | Wiley Periodicals, Inc. | 2013 | Developing a new churn prediction algorithm based on a social network analysis of the Undirected call graph and quantify the benefit of using SNA in churn prediction | Quantify the social ties, Influence propagation model, Classification of churn using logistic regression, decision trees, and random forests | Two months of CDR data | Overall lift comparison with and without using SNA predictors, Important predictor variables |
| 2 | Abd-Allah, Salah, and El-Beltagy [8] | Enhanced Customer Churn Prediction using Social Network Analysis | Proceedings of the 3rd Workshop on Data-Driven User Behavioral Modeling and Mining from Social Media - DUBMOD '14 | 2015 | Developed a method to measure social ties strength between customers via model churn as a dyadic social behavior in the Telecom network | Constructing an undirected call graph. incorporate strong social ties in an influence propagation model, and apply a machine-learning-based prediction model that combines both churn social influence and other traditional churn factors | CDR data for two months | "Call length" is probably the strongest purpose in promoting social ties |
| 3 | Olle Olle, Cai, Yuan, and Jiang [9] | Churn Influence Diffusion in a Multi-Relational Call Network | Applied Mechanics and Materials | 2015 | Quantification of Social influence propagation between customers in the social network of CCP | Built a Multi-relational call graph. Then applied Linear Threshold (LT) to model the diffusion of churners influence between customers in the social graph | 10,000 customers in China mobile network | Churn Diffusion Trends with Affinities. Churn Influence Diffusion and Probability of Retransmission in Affinity |
| 4 | Pagare, Reena Khare, Akhil [10] | Churn prediction by finding the most influential nodes in social network | 2016 International Conference on Computing, Analytics and Security Trends (CAST) | 2017 | Utilizing social network attributes for churn customers prediction | Finding Communities model then, search for the most influent nodes and model how the churners influencing is diffusions within the call graph to all other neighbor nodes | Poker social networking site of Slovakia | Executed the programs with three different approaches the first approach using Social Network Data. The second approach used Call Log Data. The third approach used customer information which performs the best accuracy |

*(continued)*

**Table 1.** (*continued*)

| No | Authors | Title | Journal | Year | Objective | Techniques | Dataset | Performance metrics |
|---|---|---|---|---|---|---|---|---|
| 5 | Kamuhanda and He [11] | A Nonnegative Matrix Factorization Approach for Multiple Local Community detection | 2018 IEEE/ACM International Conference on Advances in Social Networks Analysis and Mining (ASONAM) | 2018 | Extended the NMF for community detection method by does not use "argmax" function to assign ncies to communities and used to detect multiple local communities | Applied BFS for sampling a subgraph then use Nonnegative Matrix Factorization (NMF) to detect communities within the subgraph | Amazon, DBLP, Karateclub network, Dolphins network. | F1 score with different seed set, Conductance, community size |
| 6 | Selvaraj and Sruthi [12] | An Effective Classifier for Predicting Churn in Telecommunication | Journal of Advanced Research in Dynamical and Control Systems 11(01-special issue):221 | 2019 | The suggested model aims to find the features that highly influence of customer churn operation | Machine learning algorithms like KNN, Random Forest and XG Boost | IBM Watson dataset | F-Score, Accuracy. Fiber Optic customers with greater monthly charges attributes have higher influence for churn. XG boost classifier performs outperform the other methods |

accuracy as assessed by the F-score and high conductivity values because most of their communities are massive.

Selvaraj et al. [12] search for the attributes that highly impact the customer to churn with the assistance of machine learning techniques like KNN, Random Forest an XG Boost and to improving the accuracy of CCP. The researchers used IBM Watson Telecom dataset and show that Fiber optic customers with greater monthly charges have higher influence for churn and XGBoost classifier performs well as compared with other methods.

Comparing with a previous working in literature survey, this paper focuses on building a model of churn prediction that incorporated SNA with churn Prediction model using directed and weighted graph as opposed to most previous studies that worked in undirected networks. Also, the weight for each edge within the graph was calculated based on calls, messages, and internet traffic attributes. While most of the recent papers used only call or message in an independent manner.

We demonstrated that the ties strength and churn influence propagation model are developed to be integrated into machine learning in order to enhance the churn prediction model accuracy. The strength of social relations is calculated to describe how the influence of churn in each community within the graph is diffused through community nodes and use this policy to predict the risk of churn using linear regression as a machine learning algorithm. While most existing studies implemented the diffusion process directly in the call graph without taking into account the connection strength between community nodes and the direction of information propagation with asymmetrical ties strengths. The social strengths are calculated through the adjacency matrix of correlation attributes using the strength of the ties based on the intensity of interaction/similarity between the nodes in the community model.

Actually, When combine the SNA with churn prediction model, the computational complexity of calculation the tie strength and diffusion propagation algorithm would be totally depend on the characteristic of call/message graph topology and the nodes which represents the customers and their attribute in the social graph such as the position of node in social graph, the social interaction between the customer and all other neighbor customers in the graph.

The Telecom industry reflects different types of measurements to assess the achievement of the churn prediction model. The selection of metric entirely depends on the kind and implementation plan of the prediction model. After completing the constructing of proposed churners classification system, performance evaluators will help in evaluating model outcomes.

This paper contributes to existing literature works by suggesting a new methodology for churn prediction with the help of social network taxonomies. Based on evaluating the proposed model based on classification performance metrics such as Accuracy, Lift, Error Rate, AUC, H-Measure, Youden Index, Gini, KS, MER, MWL,... etc. Show that our churn propagation process yields the more accurate and precise tendencies of customers to churn and which are more loyal to stay and continue using the available Telcom services.

## 3   The Proposed Churn Prediction Methodology

The proposed solution concerned of how to incorporate of churn prediction model with SNA that characterize the customer behaviors in call graph, community detection method and measure the ties strength among distinctive users can be utilized with churner customer has an influence on the diffusion process and computed the total influence that ensemble for every node represents a customer in call graph. Figure 1 illustrates the proposed methodology to determine and classify Telecom customers as churners and non-churners.

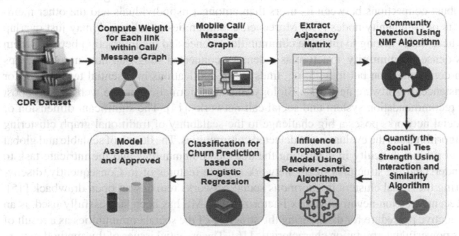

**Fig. 1.** Churn prediction based on network connection structures

### 3.1   Mobile Call/Message Graph

The fundamental structure of the mobile call graph may be characterized through the nodes that represented the customers in the Telecom provider, and a link between two nodes suggests that one or more calls have been positioned among pair of customers. In this paper, a directed and weighted call graph is constructed from the CDR data that provide the necessary information in constructing call graph e.g. Caller, Callee and the weights. The Telecom carriers generate CDR data for every call or message made to or from one in all its contacts phones and the weight in this paper is calculated for every edge of call graph which is normalized of total row sum of SMS in and out activity, call in and out activity, and internet traffic activity undertaking by each user within the CDR data. This study makes use of the term clients, customers and users Alternatively to indicate the contact in the call graph or community. The social circle or neighbors of users will also be found out as those nodes that may be reachable from the customer by means of directed ties inside the call graph. Churn happens when the customer decided to move from one Telecom service to others, users may additionally churn implicitly if they cease all use of Telecom facilities for a long time such that the likelihood that customers will return utilizing the Telecom provider someday is rare. At this point, the provider might make a recommendation to influence the client with the hope of supporting him. Nevertheless, in general, that point is too late in view that the user has already decided to churn. The effective predictive approach should no longer be the most effective in identifying churn customer but, more than that, providing long-term horizontal forecasts sufficiently in its predictions and how to reduce the churn influences on company profits.

## 3.2  Community Detection Using Non-negative Matrix Factorization (NMF)

Community discovery is a major task for revealing the general constructions in large networks. A community module or cluster is probably viewed as a group of nodes with robust connections between its users than amongst its individuals and the other members of community nodes [13]. Moreover, Communities in call graph may just overlap where nodes belong to multiple communities at once and the interaction between them is dense. **Community** structure as a densely connected subgroup now not only helps in comprehension but in addition, finds broad applications in essential techniques, for instance, in software networks, Biology, and Bioinformatics, online web sites to boost a recommendation system and social call networks [14]. The significant dimension of social networks poses a big challenge to the scalability of traditional graph clustering algorithms and the evaluation of detected communities. The dearth of scalable and global algorithms, difficulty in evaluating the detected communities and the intricate task to understand the structures of the network and the features of it. Consequently, discovering the social clusters in enormous social networks remains an open drawback [15]. Recently, a Non-negative Matrix Factorization NMF has been successfully used as an effective procedure for discovering the structure of the social communities as a result of its powerful interpretation characteristic [16]. The essential issues of this method require prior skills of the structure of the community and negative balance within the response. To address these issues, this paper use NMF with the benefit of making use of neighborhood information to identify the communities in the social network. NMF is a technique for acquiring a low-dimension description of matrices with positive elements [13]. the basis and coefficient are fundamentally based on the style of applications of interest, for instance, digital images are only matrices of non-negative digits representing pixel density. In retrieving information and extracting texts, we depend on text matrices to represent document clusters. In recommendation systems, utility matrices are illustrated the customer preferences for gadgets. This research applied NMF on the adjacency matrix of the directed and weighted call graph to partition the network into the pre-defined number of social communities. Given the adjacency matrix A of m rows and n columns with each element aij $\geq$ zero, NMF is a matrix factorization procedure which searches to scale down the adjacency matrix approximately into basis and coefficient matrices then using basis matrix to predict the communities of call graph [17]. The two non-negative matrices that obtaining after implement NMF algorithm are represented by W and H of size m rows and k columns, and k rows and n columns respectively are decided with the help of minimizing the Frobenius norm indicated in Eq. (1).

$$X \approx WH \tag{1}$$

This formulation is an NP-hard problem, each element of matrices W and H be either zero or positive. The value of k is estimated through the NMF algorithm utilizing rank function and is required to be equal or lower than the smallest value of m and n [18]. The estimated number of communities by NMF in this research is chosen to be nine social communities. Figure 2 shown the churners and non-churners count in each community of proposed call graph after implement NMF algorithm.

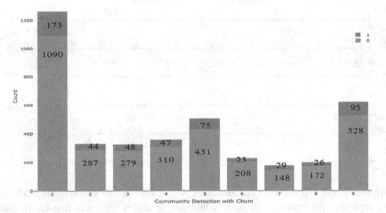

**Fig. 2.** The churners with zero value/non-churners with one value count in each community

While Fig. 3 represent the original call graph and the detected communities based on NMF technique in this paper. NMF has a complex time approach, the challenge of it can be solved via gradient descent optimization problem on a given adjacency matrix A as shown in Eq. (2).

$$\min_{H,W} \frac{1}{2} \left\{ \left\| A - WH_F^2 \right\| + \eta \left\| W_F^2 \right\| \right\} + \beta \sum_{j=1}^{n} \left\| H(:,j)_1^2 \right\| \tag{2}$$

Where W is a basis matrix, H is a coefficient matrix, $\|.\|_F^2$ is the Frobenius norm, $H(:,j)$ is the jth column vector of H, $\beta$ regularization parameter to stability the trade-off between the accuracy of the approximation and the sparseness of H, parameter $\eta$, $\beta$ are a scaling effect since a large value of $\beta$ would suppress the expression $\|W\|_F^2$ [16]. The NMF utilize L1-norm minimization and the corresponding algorithms are based on alternating non-negativity constrained least-squares and made a massive growth [19].

This paper investigates further into the NMF method as can see in the algorithm(1) and deployed it in a clustering method that used a gradient descent algorithm for optimization and clustering the call graph via data partition. The experimental results show that NMF could be used with a large matrix and it improves the performance of overlapped community detection issue, meanwhile, it can maintain the sparsity in the matrix and give better quality communities.

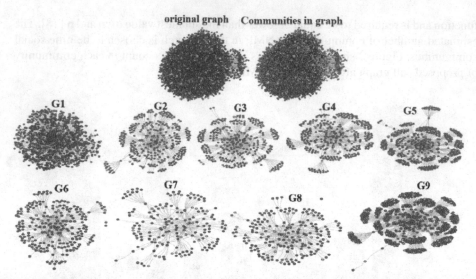

**Fig. 3.** Original graph and detected communities (G1… G9) in call graph using NMF method, pink nodes represent the churned customer while purple nodes represent non-churn customers (Color figure online)

### Algorithm(1)NMF for Social Community Extraction.

**Input:** given adjacency matrix $A \epsilon R^{m,n}$, $k \ll \min{(m,n)}$

**Output:** Coefficient matrix $H \epsilon R^{m,k}$ , Basis matrix $W \epsilon R^{k,n}$

// A*B: element-wise multiplication , AB: matrix multiplication.

1. The objective is to minimize Cost function:

$$A^2 = \left\| v^{\frac{1}{2}}(A - WH) \right\|^2 \tag{3}$$

2. For rep=1 to max repetition Do
3. $W_{old} = \text{rand}(m, k)$
4. $H_{old} = \text{rand}(k, n)$
5. $V = \|A - WH\|_F^2$     (4)
6. For i=1 to max iter Do
7. Perform specific update rule for basis:

$$\Delta H = H_{old} * \frac{W_{old}^T(A*V)}{W_{old}^T[(W_{old}H)*V]} \tag{5}$$

8. Perform specific update rule for coefficient:

$$\Delta W = W_{old} * \frac{(A*V)H_{old}^T}{[(WH_{old}^T)*V]H_{old}^T} \tag{6}$$

9. $H_{new} = \begin{bmatrix} H_{old} \\ \Delta H \end{bmatrix}$
10. $W_{new} = \begin{bmatrix} W_{old} \\ \Delta W \end{bmatrix}$
11. End for
12. End for

### 3.3 Quantification of Tie Strength

Tie strength is probably the social network concept that has attracted most researcher attention and the one that has led to the way of solving substantial contributions in social networking problems [20]. Due to the lack of a general method for measuring the strength of connection in the social network, the main objective of this paper is to bridge the gap and finds tool capable of providing quantitative and continuous measurement of the tied power between nodes in the social graph. The link strength measure involves construction the communities of call network and then model the intensity of social relations between various contacts based on CDR attributes. It's fundamental to calculate the connection strength between a social relationship based on a chosen features e.g. quantity of calls, call period, and take into count that social neighborhood circle yielded robust ties between members that point out the greater reciprocation according to the whole of social calls and the call duration [21]. In this paper, for the sake of quantifying the tie strength based on these results, it is a big challenge to model it depends on the network structural connecting such as degree, neighbors of nodes attributes and the total calling activities as a weight for each pair of nodes in a specific community. There is not exists a normal shared formal method of tie strength measure and only some procedures had been proposed with the goal of measuring the weightiness of pairwise human interactions [22]. It makes sense to model a connection strength by using a method based on the intensity of interaction and the similarity of the nodes in the community as depicted in Eq. (7).

$$t(u, v) = w(u, v) \frac{|\Gamma(u) \cap \Gamma(v)|}{\min(|\Gamma(u)|, |\Gamma(v)|)} \tag{7}$$

where w(u, v) is weight between the two nodes of u and v which representing the intensity of the interaction between nodes, $\Gamma(u)$ is the set of neighbors of a node u. It may be obvious that the knowledge provided by means of tie strength is exactly related to the number of interactions among different customers. Moreover, it is also involving the number of various contexts wherein these connections occur. In more distinct the quantification of tie-strength helps to recognize which customers are strictly correlated to each other and hence extra prone to be the influencer of different user's behavior.

### 3.4 Influence Propagation Model

Spread influence in social networking is an algorithm in which customers adapt their opinions, revise their beliefs, or exchange their habits for this reason of interactions with different users [10]. The tie intensity information used to construct an influence diffusion model described how churn influenced transforms from a churner node to his social neighbors and how much influence is preserved through a receiver in the graph. The receiver-centric algorithm is used to model the influence of the propagation process in a directed and weighted social graph. If the received node is a friendly neighbor of the sender node means it revealed a more powerful tie-intensity than every other associate then, it is usual to adopt that node might be influenced via the sender contact more than a blind node that may have social relations with it. The complete amount of preserved attraction in the constructed model is relevant to the social tie-strength that the customer has with the sender respected to the tie-strengths with all his circles [6]. The proposed of the diffusion process can describe by exploiting algorithm(2).

### Algorithm(2) Churn Influence Propagation through the social graph

**Input:** Let g: graph model, $t_{i,j}$: symbolize the tie strength between node i and jin the g, the tie strengthis directional i.e. $t_{ij} \neq t_{ji}$, $n_i$: i=1, ...,vcount of g, $n_i$ be the nodes in g.

**Output:** I: the whole churn influences accumulated at every node within the g.

1. Influence propagation for the first node in g is $I_1 = t_{1,1}$
2. Calculate the significance of the tie-intensities of all associate links of the node.Let Ni represents the set of all friends of the node then, the sum of tie-strength is depicted as $T_i = \sum_{j \in Ni} t_{i,j}$
3. The $I_{I,j}$ is the influence acquired by node i from node j is relative to the tie-strength of the edges among node i and node j to the total sum of tie-strength of all connections occurrences on node j, can be represented by using the following $I_{i,j} = I_j * \left(\frac{t_{i,j}}{T_i}\right)$,the tie influences that collected from a pair of nodes is altered because of the account of tie-strengths on each of the incoming connection.
4. The node that obtains influence will determine what portion of influence to hold. The total attraction held by node i, $I_i$ is the total of the influences discovered from all its connections and given by $I_i = \sum_{j \in Ni} I_{i,j}$
5. The receiving neighbors will again maintain a portion of the influence and user may be immediately influenced by the identical user.
6. A churner node will unfold its influence immediately to his social circle following his choice, and indirectly to all other nodes that connect with his neighbors.
7. The diffusion of the influence process goes on until it reaches the last node in g.

At the end of the diffusion model method, each node within the call graph may have a net sum of attraction that holds as a result of the entire churn behavior from it is in degree nodes. See Fig 4 to understand the churn diffusion model within a simplified call graph. At this point, node 1 is an attribute node and the tie strengths that calculated from the earlier step are marked within the links of the graph. This influence values become the influence measured that will be used as one of the attributes that will be used for churn prediction in the machine learning method. When the diffusion approach begins, node 1 will transfer all its churn effects to directly influence all its connected nodes 2, 3, and 4 and indirectly to all other nodes 5 and 6 that related to its neighbors.

Each of the targeting nodes will preserve the amount of the aggregated influence based on the relative tie-intensity with the transmitter churner node to all its circle edges. For instance, node 3 in Fig. 4 receives churner influence directly from node 1 is $I_{3,1} = I_1 \frac{t_{1,3}}{t_{1,3}+t_{2,3}+t_{3,4}+t_{3,5}}$ and indirectly from node 2 is $I_{3,2} = I_2 \frac{t_{2,3}}{t_{2,3}+t_{1,3}+t_{3,4}+t_{3,5}}$, the whole amount of churn influence at node 3 is then $I_3 = I_{3,2} + I_{3,1}$. The receiving nodes will again preserve an amount of the churn influence and then transform it to all its neighbors. The propagation of churn goes on until attaining the last node within the graph. Subsequently, each node of the graph has a complete influence value that will be used as a

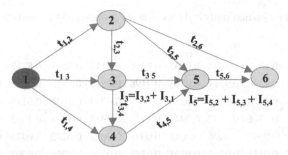

**Fig. 4.** Churn influence propagation model

predictor for churn. The end-stage in the churn prediction model utilizing SNA taxonomy is to integrate the churn influence event into an ordinary machine learning algorithm to classify the churners users. Classification algorithm such as logistic regression is used with target label variable to calculate the tendency for churn of distinct customers. LR model applies when the response variable can be assumed to be a continuous variable or to be normally distributed. It's a simple statistical model because there will be an exact solution for the regression parameters and can solve algebraically.

## 4 Experimental Analysis and Results

Robust practical setup and use of statistical tests and appropriate performance measures are crucial to figuring a correct conclusion. The aim of research is to construct an experimental illustration of the model performance using relational classifier for classification churner customers in mobile telco company founded on network content and quantify the intensity of friendly links among different users using different features and how they churn influences are spread directly from churn node to all it is neighbor nodes and indirectly to all nodes that construct a subgraphs that contained these nodes. The selection of statistical achievement measures should depend on the scope of the churn model and consider how to handle them. Table 2 represents the results of proposed churners prediction model from different assessment metrics e.g. accuracy, F-Score, and AUC in the suggested churn prediction system [22].

### 4.1 F-Score

Precision is invaluable for assessing the performance of data mining classifiers, but it surely leaves out some facts and for that reason will also be complicated. The recall is a portion of the true optimistic predictions to total positive observations in the dataset. Compute the percent of churn rate that appropriately categorized as churn/non-churn. The prediction models that have a low recall means it miss-classifies a great amount of the positive cases. The F-Score can be outlined as in Eq. 8.

$$F - Score = 2.(Precision * Recall)/(Precision + Recall) \qquad (8)$$

**Table 2.** Results of evaluation statistical measures for proposed churn prediction methodology

| Community | 1 | 2 | 3 | 4 | 5 | 6 | 7 | 8 | 9 | Average |
|---|---|---|---|---|---|---|---|---|---|---|
| ACC | 0.8651 | 0.8769 | 0.8594 | 0.8732 | 0.8416 | 0.9111 | 0.8529 | 0.8718 | 0.8468 | 0.87 |
| F-score | 0.9254 | 0.9344 | 0.9244 | 0.9323 | 0.9198 | 0.9535 | 0.8852 | 0.9315 | 0.917 | 0.92 |
| AUC | 0.5303 | 0.5088 | 0.6036 | 0.5253 | 0.5275 | 0.5244 | 0.7635 | 0.6176 | 0.5263 | 0.57 |
| Thresh | 0.1108 | 0.1346 | 0.1282 | 0.0943 | 0.1434 | 0.0954 | 0.1439 | 0.1063 | 0.1495 | 0.12 |
| H- | 0.0011 | 0.0044 | 0.0096 | 0.0041 | 0.0093 | 0.0057 | 0.0222 | 0.0075 | 0.0395 | 0.01 |
| Youden | 0.0046 | 0.0175 | 0.0364 | 0.0161 | 0.0349 | 0.0244 | 0.0966 | 0.0294 | 0.0526 | 0.03 |
| Gini | 0.0324 | 0.0176 | 0.1818 | 0.0162 | 0.0550 | 0.0488 | 0.4620 | 0.2352 | 0.0526 | 0.12 |
| KS | 0.0046 | 0.0175 | 0.0364 | 0.0161 | 0.0349 | 0.0244 | 0.0966 | 0.0294 | 0.0526 | 0.03 |
| MER | 0.1310 | 0.1231 | 0.1406 | 0.1268 | 0.1485 | 0.0889 | 0.1471 | 0.1282 | 0.1452 | 0.13 |
| MWL | 0.2266 | 0.2121 | 0.2417 | 0.2178 | 0.2441 | 0.1580 | 0.2266 | 0.2170 | 0.2458 | 0.22 |
| ER | 0.1389 | 0.1385 | 0.1250 | 0.1408 | 0.1386 | 0.1111 | 0.1765 | 0.1282 | 0.1613 | 0.14 |
| Lift | 1.4820 | 1.2500 | 1.1100 | 1.1100 | 2.6700 | 0.8330 | 0.6670 | 4.0000 | 0.5260 | 1.52 |

## 4.2 The Area Under Receiver Operating (AUC)

Measures the subject underneath ROC curve, the diagonal line represents a random process, it has an AUC of 0.5, thus the AUC of a reputable churn classifier should be a lot higher, preferably virtually 1, as a worth of 1 represents ultimate classifier. The field-specific by means of AUC represents the chance that a random pair of churning and non-churning customers are properly identified, i.e. A positive instance receives a greater rating than a negative instance.

## 4.3 Lift Curve

The effectivity of the prediction model is expressed in the lift curve, which emphases on the group of customers with the highest hazard to attrite the service provider when a designated fraction of subscribers used to be contacted. This is equal to the ratio between the sensitivity and the ratio of predicted churners after applying the churn model to the testing dataset. Formula 9 represents the lift value [23].

$$lift = \frac{precision}{p/(p+N)} \tag{9}$$

## 4.4 Receiver Operating Characteristic Curve (ROC)

The chart is a depiction of the relations between the benefits, costs, a plot in two-dimensional space of x- and y-axes in linear scale and used to summarize the trade-off between recall and 1-specificity [24].

## 4.5  Kolmogorov-Smirnoff Statistic (KS)

Measure the amount of separation between desirable and undesirable distributions. The KS statistic test gives the maximum distance between the ROC curve and the sloping at a specific cut-off point. In most prediction models, the KS test falls in the range of zero and one, the higher value means better model in separating the positive from negative classes.

## 4.6  Accuracy

Count the correct predictions accomplished by the classification model over all kinds of predictions made. Overall, how often the classifier model is correct.

## 4.7  Cost

Many of the above metrics attempt to take a balanced view between FP and FN. A principled method to acquire this is to introduce the suggestion of misclassification expenses. Let c in [0, 1] denote the 'price' of misclassifying a category zero object as category one (FP), and $1 - c$ the fee of misclassifying a category one object as category zero (FN) based on the calculated confusion matrix. It's implicitly assumed that the two misclassification expenses sum is 1. Minimal Error Rate MER, Minimum Weighted Loss (MWL). MER is a targeted case of MWL for $c = 0.5$ [23].

## 4.8  Gini Coefficient

The Gini coefficient also relates to assessing classifier models. It is directly related to the area under the ROC curve and computed by formula $2 * AUC - 1$.

## 4.9  H-measure

Some researchers have been proven that the problem of the AUC is that it depends totally on the use of data mining method and differed based on the classification method. H-measure is successfully overcoming the variance of AUC, so it captures the performance advantage of AUC but not its flaw i.e. incoherent and potentially misleading yields when the ROC curve is cross [25].

Most Telcom carriers have incentives to attach users inside their networks like free calls or lowered costs to inspire associates of the user to become a member of their network service providers. The percentage of user calls inside a mobile Telecom network and outside it has an excessive social intention. If the customer makes most calls external the network, it's probable that the person will recall switching to another supplier sooner or later, to save fees. The CDR dataset in this study was limited and didn't incorporate any demographic or popular evidence or any learning about customer network charges and abandoned calls. It contains voice call; SMS messages and internet activities were available for use in the prediction model. A directed and weighted call graph using CDR dataset was created, community detection using NMF method used to estimate the number of communities and to partition the graph into many communities with modularity equal 0.068 so it can be more accurate to implement and follow up the prediction strategy in each community than the whole graph. To be precise, identifying communities is an unspecified problem, there are no universal protocols on key components, such as the definition of the community itself, nor on other critical issues, such as validation of algorithms and comparison of their outputs. The social tie-strength was calculated for all the edges between nodes. Then, churn diffusion model through nodes of the graph was implemented based on tie-strength of each link between churner nodes and its neighbors in the network.

Table 2 figured the Performance statistics calculated for the final model based on the confusion matrix of a churn predictive model using a logistic regression model. Classification learner based on network connections and community detection methodology has a strong predictive ability with accuracy value for all communities in the graph between 84% to 91%. We observed a significant variation in lift performance based on how the dataset was divided into many communities and split for training and testing purposes. The model was run for many data partitioning and the average lift for the first ten was computed to be in the range of 0.5 to 4. The content of Table 1 of performance metric are drawing as bar plot to give more illustration of performance values, see Fig. 5 and Fig. 6 represents the total average of each evaluation metric to give more insights about the overall call graph.

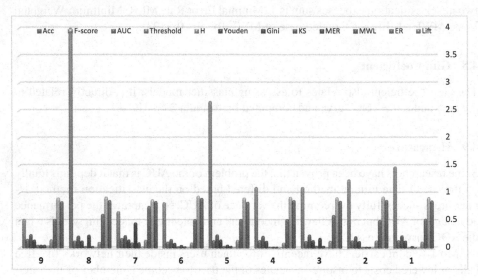

**Fig. 5.** Evaluation metric for the churn prediction model

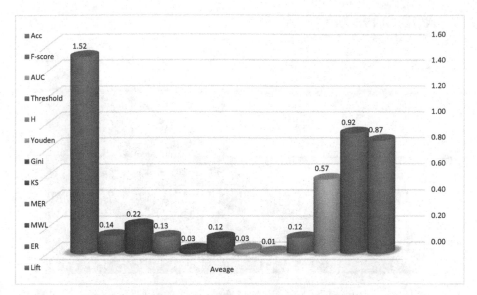

**Fig. 6.** Average of evaluation metrics for churn prediction model

From Fig. 7 the seventh community outperforms the other communities in terms of AUC metric with value 0.7635 due to its position close to the upper left. While the worst case is the second community with AUC is 0.5088 since it is very close to the baseline and not has a concave region, and the average AUC overall churn labels were 0.57. The striking thing in the table is that the AUC and H are not simply monotonically related; whereas the AUC is large H is substantially small, and many of ROC curves are shown overlaying.

AUC summarizes the overall location of the entire ROC curve. It is of great interest since it has a meaningful interpretation. it can be interpreted as the average value of sensitivity for all the possible values of specificity. The maximum AUC = 1 means that the diagnostic test is perfect in the differentiation between the churn and non-churn customers. This occurs when the distribution of test results for the churn and non-churn do not overlap. AUC = 0.5 means the chance discrimination that curve located on the diagonal line in ROC space. The minimum AUC should be considered a chance level i.e. AUC = 0.5 while AUC = 0 means test incorrectly classify all subjects with churn as negative and all subjects with non-churn as positive that is extremely unlikely to happen in practical of the churn prediction system.

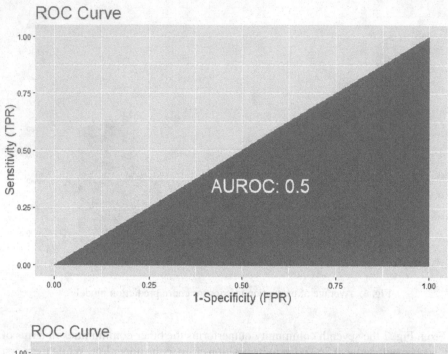

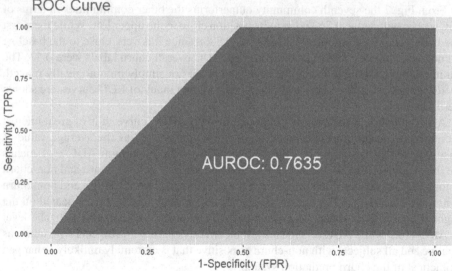

**Fig. 7.** ROC curves (seventh community in the bottom) and (second community in the top)

## 5  Conclusion

The main contribution of this study is to determine the strength of social ties between users for each identified community based on CDR attributes and then applies a model to propagate the churn influences on the call graph to decide the net aggregated influences from the churner node. We use this effect in the logistic regression method to calculate the tendency of churn for individual users. The churn prediction model performance can be increased by considering the relationship between the users during the construction of the social graph of Telecom. In this work, we use the CDR attributes that contain information about phone calls and SMS messages to model the relational strength of the social communication between edges in the graph. NMF method allowed to find the ideal algorithm of building the communities of the network regarding links and weights, and to decide the actual attributes from CDR data about calls and messages that are essential to achieve the optimal achievement. It observed that the strength of the tie is close regarding the number of interactions between the members of the network. Additionally, the quantity can be related to the communities where these connections arise. Computing the social tie-strength play a vital role in implementing of churn propagation model that aims to decide the portion of influence of specific user is given from churner users.

As future work, it may be a plan to investigate how the knowledge offered with the help of tie-strength approach may also be exploited to address social network issues such as link prediction and clusters discovery. In addition, as a result of the dearth of study on the outcome of call structure on typical achievement and community detections, then as future work, they would moreover help in understanding human behavior in the case of social ties and churn.

# References

1. Modani, N., Dey, K., Gupta, R., Godbole, S.: CDR analysis based telco churn prediction and customer behavior insights: a case study. In: Lin, X., Manolopoulos, Y., Srivastava, D., Huang, G. (eds.) WISE 2013. LNCS, vol. 8181, pp. 256–269. Springer, Heidelberg (2013). https://doi.org/10.1007/978-3-642-41154-0_19
2. Amer, M.S.: Social network analysis framework in telecom. Int. J. Syst. Appl. Eng. Dev. **9**, 201–205 (2015)
3. Saravanan, M., Vijay Raajaa, G.S.: A graph-based churn prediction model for mobile telecom networks. In: Zhou, S., Zhang, S., Karypis, G. (eds.) ADMA 2012. LNCS (LNAI), vol. 7713, pp. 367–382. Springer, Heidelberg (2012). https://doi.org/10.1007/978-3-642-35527-1_31
4. Shobha, G.: Social network analysis for churn prediction in telecom data. Int. J. Comput. Commun. Technol. **3**(6), 128–135 (2012)
5. Gamulin, N., Štular, M., Tomažič, S.: Impact of social network to churn in mobile network. Automatika **56**(3), 252–261 (2015)
6. Rijnen, M.: Predicting Churn using Hybrid Supervised/Unsupervised Models (2018)
7. Phadke, C., Uzunalioglu, H., Mendiratta, V.B., Kushnir, D., Doran, D.: Prediction of subscriber churn using social network analysis. Bell Labs Tech. J. **17**(4), 63–75 (2013)
8. Abd-Allah, M.N., Salah, A., El-Beltagy, S.R.: Enhanced customer churn prediction using social network analysis. In: Proceedings of the 3rd Workshop on Data-Driven User Behavioral Modeling and Mining from Social Media - DUBMOD 2014, January 2015, pp. 11–12 (2015)
9. Olle Olle, G., Cai, S.Q., Yuan, Q., Jiang, S.M.: Churn influence diffusion in a multi-relational call network. Appl. Mech. Mater. **719–720**, 886–896 (2015)
10. Pagare, R., Khare, A.: Churn prediction by finding most influential nodes in the social network. In: 2016 International Conference on Computing, Analytics and Security Trends (CAST), pp. 68–71 (2017)
11. Kamuhanda, D., He, K.: A nonnegative matrix factorization approach for multiple local community detection. In: 2018 IEEE/ACM International Conference on Advances in Social Networks Analysis and Mining (ASONAM), pp. 642–649, November 2018

12. Selvaraj, S., Sruthi, M.: An effective classifier for predicting churn in telecommunication. J. Adv. Res. Dyn. Control Syst. **11**, 10 (2019). (01-special issue)221
13. White, K.: Social Networks Analysis. The Sage Dictionary of Health and Society (2012)
14. Qu, Y., et al.: Exploring community structure of software Call Graph and its applications in class cohesion measurement. J. Syst. Softw. **108**, 193–210 (2015)
15. Du, R., Drake, B., Park, H.: Hybrid clustering based on content and connection structure using joint nonnegative matrix factorization. J. Global Optim. **74**(4), 861–877 (2017). https://doi.org/10.1007/s10898-017-0578-x
16. Zhu, G.: Nonnegative Matrix Factorization (NMF) with Heteroscedastic Uncertainties and Missing Data (2016)
17. Schachtner, R., Lang, E.W., Pöppel, G.: Extensions of Non-negative Matrix Die vorliegende Dissertationsschrift entstand während einer dreijährigen Zusammenarbeit mit der Firma Infineon Technologies AG Regensburg. Wissenschaftliche Betreuer (2010)
18. Kamuhanda, D., He, K.: A nonnegative matrix factorization approach for multiple local community detection. In: Proceedings of 2018 IEEE/ACM International Conference on Advances in Social Networks Analysis and Mining, ASONAM 2018, pp. 642–649, July 2018
19. Kim, H., Park, H.: Sparse non-negative matrix factorizations via alternating non-negativity-constrained least squares for microarray data analysis. Bioinformatics **23**(12), 1495–1502 (2007)
20. Onnela, J.-P., et al.: Structure and tie strengths in mobile communication networks. Proc. Natl. Acad. Sci. **104**(18), 7332–7336 (2007)
21. Abd-Allah, M.N., El-Beltagy, S.R., Salah, A.: DyadChurn: customer churn prediction using strong social ties. In: 2017 Proceedings of the 21st International Database Engineering & Applications Symposium - IDEAS 2017, pp. 253–263, January 2018
22. Dino Pedreschi, F.G.: Social Network Dynamics (2015)
23. Óskarsdóttir, M., Bravo, C., Verbeke, W., Sarraute, C., Baesens, B., Vanthienen, J.: Social network analytics for churn prediction in telco: model building, evaluation, and network architecture. Expert Syst. Appl. **85**, 204–220 (2017)
24. Halibas, A.S., Cherian Matthew, A., Pillai, I.G., Harold Reazol, J., Delvo, E.G., Bonachita Reazol, L.: Determining the intervening effects of exploratory data analysis and feature engineering in telecoms customer churn modelling. In: 2019 4th MEC International Conference on Big Data and Smart City (ICBDSC), pp. 1–7 (2019)
25. Hand, D.J.: Measuring classifier performance: a coherent alternative to the area under the ROC curve. Mach. Learn. **77**(1), 103–123 (2009)

# The Impact of Rule Evaluation Metrics as a Conflict Resolution Strategy

Nabeel H. Al-A'araji[1]([⊠]), Safaa O. Al-Mamory[2]([⊠]), and Ali H. Al-Shakarchi[3]([⊠])

[1] Ministry of Higher Education, Baghdad, Iraq
nhkaghed@itnet.uobabylon.eda.iq
[2] Collage of Business Informatics, University of Information Technology and Communications, Baghdad, Iraq
salmamory@uoitc.edu.iq
[3] Collage of Information Technology, University of Babylon, Hillah, Iraq
ali.al-shakarchi@itnet.uobabylon.edu.iq

**Abstract.** It is crucial for each rule induced via machine learning algorithm is to be associated with a numerical value(s) which can reflect its properties like accuracy, coverage, likelihood. The accumulation of these properties is the so-called evaluation metrics. These metrics are important for both rule induction systems (for stopping rules generation) and rule classification systems (for solving rules conflict). This paper describes the most important of both statistical and empirical rule evaluation metrics. Thereafter, the paper presents an approach that utilizes and shows the impact of these metrics as a rule conflict resolution strategy during classification tasks when combining two heterogeneous classifiers (Naïve Bayes (henceforth, NB) and decision tree J48). To accomplish this goal, authors have extracted rule-set from J48 and presented a new method for rule construction from NB. Experiments have been conducted on (WBC, Vote, and Diabetes) datasets. The experimental results show that different evaluation metrics have a different impact on the classification accuracy when used as a conflict resolution strategy.

**Keywords:** Rule evaluation metrics · Rules conflict resolution · Classification · Rule extraction

## 1 Introduction

Many rule induction systems have been developed in the past decades [1, 2]. These systems utilize training dataset to generate decision rules (henceforth, rule-set). The induced rule-set is then exploited by a classification system to classify the unseen pattern. Therefore, it is essential that rule-set reveals an important feature of the classification system that is high predictability or reliability [3].

In fact, the learning algorithm could induce rule-set in two different approaches that are ordered or unordered rule-set. In the case that rule-set is generated via the first approach, the decision of a classification system is done in a straightforward way: the classifier searches rule-set and picks the first rule that matches the given pattern. On the other hand, if the rule-set is induced under the second approach, the classifier will face three different cases [4]:

© Springer Nature Switzerland AG 2020
A. M. Al-Bakry et al. (Eds.): NTICT 2020, CCIS 1183, pp. 105–127, 2020.
https://doi.org/10.1007/978-3-030-55340-1_8

1. In the case, that tested pattern satisfies one or more rules, and these rules refer to the same class, then the pattern will be assigned to rule(s) to fire referred class.
2. If the pattern is not covered by any rule in the rule-set, then the classifier either tells the user that it is unable to make the decision on appropriate class, or the pattern has been assigned to default class depending on the training dataset, or invoke other similar technique.
3. The major problem arises when the pattern is covered by rules that are relevant to different classes. This situation is referred to as rules conflict and must be handled by an appropriate strategy to pick a true class (which is our concern in this paper).

There is no doubt that rule evaluation metrics are necessary for both rule induction and classification tasks. In the former task, these metrics are needed as a criterion of rule generalization and specification. While in the latter task, metrics are attached to each rule to be used for resolving conflict when many rules with different classes are triggered by the same pattern [3].

This paper describes the most attractive of both statistical and empirical rule evaluation metrics. Moreover, the paper presents a new approach that utilizes and shows the impact of these metrics as a rule conflict resolution strategy during classification tasks when combining two different (heterogeneous) classifiers named Naïve Bayes (henceforth, NB) and decision tree J48 classifiers. Since the previous classifiers are not Rule-Based Classifiers (henceforth, RBC), this implies to construct rule-set of both classifiers to be used as RBC. For these reasons, rule-set from J48 has been constructed and a new method for rule construction from NB has been presented. Our experiments are carried out using three different datasets (WBC, Vote, and Diabetes). The experimental results show that different evaluation metrics have a different impact on the classification accuracy when used as a conflict resolution strategy.

This paper is organized as follows: the next Section gives brief background about RBC, shows the motivations to present this paper, and presents an overview of related work. Section 3 describes in detail two different types of evaluation metrics: statistical and empirical. Thereafter, Sect. 4 gives a short overview of NB and J48 classifier as well as describing the method of constructing rule-set from J48 and presenting a new method to construct rule-set from NB classifier. The methodology for resolving a conflict between rules has been described in Sect. 5. Section 6 will explain experimental and obtained results. Finally, the authors conclude the paper.

## 2 Background

### 2.1 Rule-Based Classifiers (RBC)

A simple way of demonstrating information or knowledge is via rules. RBC explodes a set of IF-THEN rules (i.e. rule-set) for classification tasks. IF-THEN rules are expressed as follow:

IF condition THEN conclusion

The left-hand side (IF part) is known as the rule antecedent (including one or more attributes as a condition), while the right-hand side (THEN part) is the rule consequent (express the class label) [5].

## 2.2  Motivation

A rule covers a record if its antecedent is met. The rule coverage is expressed by the proposition of records covered by a rule as shown in Eq. 1 below:

$$Coverage = \frac{\textbf{Number of records satisfy by rule}}{\textbf{Total number of records}} \tag{1}$$

According to the above equation, an ideal RBC must have two important properties for their rule-set that is: **Mutually exclusive** and **Exhaustive**. A rule-set is said to be mutually exclusive if there are no two rules triggered by the same pattern in the dataset. On the other hand, exhaustive rule-set ensure that there is only one rule for every combination of feature values (i.e. there exists a rule for every part of feature space) [6].

The two properties mentioned above ensure that every example is covered exactly by one rule, however, this is an ideal case and not all rule-based classifiers have this property. Violation of exhaustive property may cause some examples in a dataset not to be triggered by any rule, on the other hand, if mutually exclusive property is not met, several rules may be triggered by the same example and if the triggered rules produce a different class, a conflict will arise. However, if a conflict occurs, a strategy for resolution should be provided (this is our concern in this paper).

The above problem motivates researchers to find an exit to resolve conflict rules. There are many suggested rule conflict resolution strategies in the literature. Each of them utilizes different aspects to resolve conflict. For example, some conflict resolution strategies focus on generating ordered rule-set in which the classification system will fire the first trigger rule, other strategies have utilized the size of rule antecedent, that is the larger rule antecedent, the preferred fired rule. Another attractive strategy is to use evaluation metrics during rule generation and pruning processes.

Due to the above motivation, this paper, firstly, presents a description of the most important rule evaluation metrics with two concepts: statistical and empirical metrics. Thereafter the paper employs these metrics to present an approach that has been used to solve a conflict that may arise during classification tasks when combining two different (heterogeneous) classifiers named NB and J48.

Since the proposed work has dealt with heterogeneous classifiers (i.e. NB & J48) which are not RBCs, this implies to construct rule-set of both classifiers to be used as RBC. For these reasons, a rule-set from J48 has been extracted and a new method to construct rule-set from NB has been presented. Our experiments are conducted on three different datasets (WBC, Vote, and Diabetes). Obtained results show that different rule evaluation metrics have a different impact on classification system accuracy when used as a conflict resolution strategy.

## 2.3  Related Work

Rule evaluation metrics and conflict resolution strategies are attracting extensive attention from researchers in the literature. Rule evaluation metrics collected in the surveys [7–11]. What is attractive in these surveys is that each work focuses on a distinct subset of the rule evaluation metrics.

Janssen et al. present research on parameterized quality measures and the procedure of data-driven selection of their parameters [8]. Papers by Yao et al. [11] and Lavrac et al. [9] define a set of evaluation metrics using the probability concept.

Tony Lindgren presents a survey of methods that have been used to resolve conflict (9 methods). He also introduces a novel method called Recursive Induction and concludes that his method outperforms other methods [12].

Work done by Łukasz Wróbel et al. presents results about the efficiency of the rule evaluation metrics taking into their consideration the stages of rule induction systems (i.e. rule growing and pruning) as well as conflict resolution. They focus on eight evaluation metrics and use sequential covering rule induction algorithm in their work [11].

Zaixiang Huang et al. [13] propose an approach called Associative Classification with Bayes (AC-Bayes) that addresses rules conflict. Whenever rule conflicts occur, they collect those records covered by rules to generate a new training set. Thereafter, they compute the posterior probabilities for each class on test instance. The winning instance will be assigned to the class value with maximum posterior probability.

Aijun An et al. describe in their work a set of statistical and empirical evaluation metrics. They present experimental comparisons of these metrics on a number of machine learning datasets. Also, their work defines meta-behavior metrics which concluded from their experiments shows the relationship between rule evaluation metrics and the characteristics of a dataset [3].

Ivan Bruha et al. present two different approaches to define rule evaluation metrics: empirical and statistical. They introduced a practical formula of rule metrics used in induction systems and compared the characteristic of these formulas [4].

Again, Aijun An et al. enlarge their previous set of statistical and empirical rule quality formulas which they tested earlier on a number of standard machine learning datasets. In this paper, they did an extensive comparison between metrics in this and previous work [14].

Almost all previous works focus on rule evaluation metrics either in induction or in pruning tasks, however, some work concentrates on rule evaluation metrics to solve a conflict in just one classification model. Our proposed work will show the impact of rule evaluation metrics as a conflict resolution strategy when combining two heterogeneous (not rule-based) classifiers. And this approach could be extended to an ensemble system with more than two classifiers.

## 3   Rule Evaluation Metrics

Rule evaluation metrics could be obtained via extensive analysis of the relationship between a rule R and the class C. What reflects this relationship is the so-called contingency table (sometimes called two-way frequency table) which can be depicted by $2 \times 2$ table [15]. The contingency table is a vital numerical matrix commonly used in machine learning tasks [16] to present categorical data in the form of frequency counts. Table 1 below shows the contingency table:

**Table 1.** Contingency table with absolute frequency

|  | Class C | Not class C |  |
| --- | --- | --- | --- |
| Covered by rule R | $n_{rc}$ | $n_{r\bar{c}}$ | $n_r$ |
| Not covered by rule R | $n_{\bar{r}c}$ | $n_{\bar{r}\bar{c}}$ | $n_{\bar{r}}$ |
|  | $n_c$ | $n_{\bar{c}}$ | $N$ |

where: $n_{rc}$: is the number of patterns in training data covered by rule R and correspond to class C; $n_{r\bar{c}}$: is the number of patterns in training data covered by rule R but not correspond to class C; and so forth to the rest. $n_r$, $n_{\bar{r}}$, $n_c$, $n_{\bar{c}}$: are marginal totals, that is: $n_r = n_{rc} + n_{r\bar{c}}$ refer to number of patterns covered by rule R. $N$: total number of training patterns.

We can write the contingency table in a form of relative frequency rather than absolute frequency as shown in Table 2.

**Table 2.** Contingency table with relative frequency

|  | Class C | Not class C |  |
| --- | --- | --- | --- |
| Covered by rule R | $f_{rc}$ | $f_{r\bar{c}}$ | $f_r$ |
| Not covered by rule R | $f_{\bar{r}c}$ | $f_{\bar{r}\bar{c}}$ | $f_{\bar{r}}$ |
|  | $f_c$ | $f_{\bar{c}}$ | 1 |

where $f_{rc} = \frac{n_{rc}}{N}$, $f_{\bar{r}c} = \frac{n_{\bar{r}c}}{N}$ and so forth.

## 3.1 Empirical Evaluation Metrics

Bruha et al. [4] see that some rule evaluation metrics have special behaviour to evaluate a rule, they called these metrics as **Empirical formulas** since they are backed up via surmise and not by statistical or information theories.

Before introducing the most important empirical evaluation metrics, the paper will present two characteristics of a rule used to derive empirical evaluation metrics (which also could be used and rule evaluation metrics). The first one is **Consistency** (also called accuracy), while the second is **Coverage** (also called completeness) of the rule R. So by utilizing components of a contingency table, the consistency is defined in Eq. 2 as follow:

$$Cons(R) = \frac{n_{rc}}{n_r} \tag{2}$$

The coverage of rule R in Eq. 3 below is derived as follows:

$$Cover(R) = \frac{n_{rc}}{n_c} \tag{3}$$

The above two metrics are crucial indicators of the rule's reliability. Two things must be taken into consideration when intending to use these metrics as evaluation metrics: the first one is about coverage metric, here we said that rules which cover positive patterns may also cover negative ones. The second is about the accuracy, using only this metric may produce rules that cover just a few patterns in the dataset, and this will lead to poor predictive performance because rules may overfit the dataset. It is beneficial to use both consistency and coverage metrics to derive new powerful metrics.

**Weighted Sum of Consistency and Coverage**
Michalski and Ryszard [1] have proposed a new evaluation metric that makes using a weighted sum of both consistency and coverage as shown in Eq. 4:

$$Qws(R) = w1 \times cons(R) + w2 \times cove(R) \tag{4}$$

where w1, w2 are weights defined by a user, and their values confined between (0, 1) and summed to 1. This metric has been used in YAILS learning system [17]. This system defined weights as follows:

$$W1 = 0.5 + \frac{1}{4}cons(R) \text{ and } W2 = 0.5 - \frac{1}{4} \times cons(R)$$

The weights above are influenced by the consistency that is the larger consistency, the more influence rule by consistency. One can replace the consistency with the coverage metric.

**Product of Consistency and Coverage**
Product of both consistency and coverage proposed to be used as an evaluation metric by [18] is expressed in Eq. 5 below:

$$Q_{prod}(R) = cons(R) \times f(cover(R)) \tag{5}$$

where $f$ is a function, and authors have conducted extensive experiments where they find that the best function of $f$ is the exponential function as $= e^{x-1}$. They also find that $f$ reduces the effect on coverage differences on rule metric.

### 3.2  Statistical Evaluation Metrics

Empirical evaluation metrics have been seen to be easy to understand and their results are quite comparative, but they do not depend on any theoretical aspect. Nevertheless, they could be used as plausible rule evaluation metrics. Bruha, Ivan and Kockova, S [4], Aijun An and Nick Cercone [3] define three groups of statistical evaluation metrics. All of them depend statistically on the contingency table.

**Measures of Association**
This type of measure is obtained via the relation between classification for rows and columns in the contingency table.

- **Pearson $x^2$ Statistic:** The assumption that $x^2$ based on is as follows: whenever the classification of columns is independent on that on rows, the frequencies in the cells of the contingency table should be proportional to the marginal totals. $x^2$ is obtained in Eq. 6 as follows:

$$x^2 = \sum \frac{\left(n_{rc} - \frac{n_{r}n_{c}}{N}\right)^2}{\frac{n_{r}n_{c}}{N}} \tag{6}$$

If the correlation between class C and rule R is high, then the $x^2$ shows the small value.

- **G2 Likelihood Ratio Statistic:** This metric measures the range of the detected frequency distribution of patterns among classes satisfying rule R and predictable frequency distribution of the same number of the patterns under consideration that rule R selects patterns randomly. This metric is used in CN2 induction system [2]. Using the contingency table, G2 could be obtained as shown Eq. 7 below:

$$G2 = 2\left(\frac{n_{rc}}{n_r}log\frac{n_{rc}N}{n_r n_c} + \frac{n_{r\bar{c}}}{n_r}log\frac{n_{r\bar{c}}N}{n_r n_{\bar{c}}}\right) \tag{7}$$

where the logarithm is of base e. The lower G2 values indicate that the association between two distributions is due to chance.

**Measures of Agreement**
Evaluation metrics obtained under this category used contingency table with relative frequencies and focus only to diagonal elements [4, 14].

- **Cohen's Formula:** The summation of main diagonal using of relative frequencies contingency table: $f_{rc} + f_{\bar{r}\bar{c}}$ will measure actual agreement, while the chance agreements obtained as: $f_r f_c + f_{\bar{r}} f_{\bar{c}}$. Cohen [19] suggests a new rule evaluation metric expressed in Eq. 8 that compares the actual and chance agreements as follows:

$$Q_{Cohen} = \frac{f_{rc} + f_{\bar{r}\bar{c}} - (f_r f_c + f_{\bar{r}} f_{\bar{c}})}{1 - (f_r f_c + f_{\bar{r}} f_{\bar{c}})} \tag{8}$$

- **Coleman's Formula:** Coleman [15] presents a rule evaluation metric which computes the agreement association between the first column and any row in the contingency table. This metric is also obtained by normalizing the difference between the actual and chance agreement as in Eq. 9 below:

$$Q_{coleman} = \frac{f_{rc} - f_r f_c}{f_r - f_r f_c} \tag{9}$$

**C1 and C2 Formulas:** Bruha [4] conducts an extensive analysis of Coleman's metric and finds that it does not compromise the coverage. Also, he concludes that Cohen's metric is more dependent on coverage. Therefore, he modifies Coleman's metric in two different way and yields C1 (Eq. 10) and C2 (Eq. 11) metrics as follows:

$$Q_{C1} = Q_{coleman} \times \frac{2 + Q_{Cohen}}{3} \tag{10}$$

$$Q_{C2} = Q_{coleman} \times \frac{1 + cover(R)}{2} \tag{11}$$

where the coefficients 2, 3 and 1, 2 are used for the normalization purpose.

**Measure of Information**
A statistical measure of information could also be used as a rule evaluation metric. The idea is as follows: given class C, the amount of information needed to classify pattern into class C whose prior probability is P(C) is defined as $(-\log P(C))$ [20]. Here the log is of base 2. Kononenko and Bratko [20] present a new evaluation metric utilizing the previous idea, and they gave the name of this metric as an *information score* which can be expressed Eq. 12 as follows:

$$Q_{IS} = -\log\frac{n_c}{N} + \log\frac{n_{rc}}{n_r} \tag{12}$$

**Measure of Logical Sufficiency**
Another statistical measure of rule quality is logical sufficiency. Given rule R and class C the logical sufficiency of rule R with respect to class C is computed as shown in Eq. 3 below:

$$Q_{LS} = \frac{P(R|C)}{P(R|C^-)} \tag{13}$$

where P denotes the probability. The $Q_{LS}$ could be written in the sense of contingency table as in Eq. 14:

$$Q_{LS} = \frac{\frac{n_{rc}}{n_c}}{\frac{n_{r\bar{c}}}{n_{\bar{c}}}} \tag{14}$$

**Measure of Discrimination**
A measure of discrimination could be considered as a statistical evaluation metric for a rule. It is used in ELEM2 [21]. This formula was inspired by the query formula used in probability-based information retrieval. If we project this formula into rule-set to evaluate a rule R, we can consider rule R as a query in a retrieval system, positive patterns of class C as relevant documents, and negative patterns are non-relevant documents. Then

the following formula (Eq. 15) could be used as a measure of the extent to which the rule R can recognize between the positive and negative patterns of class C [14]:

$$Q_{MD} - log \frac{P(R|C)(1 - P(R|\overline{C}))}{P(R|\overline{C})(1 - P(R|C))} \tag{15}$$

where P denotes probability. The above formula can be recomputed via contingency table as in Eq. 16 below:

$$Q_{MD} = \frac{\frac{n_{rc}}{n_{\overline{r}c}}}{\frac{n_{r\overline{c}}}{n_{\overline{r}\overline{c}}}} \tag{16}$$

**Measures Depend on Coverage**

There are two important measures for rule evaluation that takes into account the coverage of rule R when producing the metric to evaluate rules. The first one is the *Laplace* metric which is obtained in as in Eq. 17 below [22]:

$$\textbf{Laplace} = \frac{f_+ + 1}{n + k} \tag{17}$$

while the second is *M-estimate* (Eq. 18) which is obtained as follows:

$$\textbf{M - estimate} - \frac{f_+ + kp_+}{n + k} \tag{18}$$

where $n$ is number of patterns covered by rule R, $f_+$ is the number of positive covered by rule R, $k$ is a total number of classes, and $p_+$ is a prior probability of positive class. Note that when $p_+$ is equal to $1/k$, the Laplace and M-estimate will be equivalent.

## 4 Rules Construction

Many works in the literature for rule constructing have been proposed, but the effort is ongoing to find out more accurate and comprehensible methods. Manomita Chakraborty et al. [23] propose an algorithm named Reverse Engineering Recursive Rule Extraction (RE-Re-RX) to extract rules from NN with mixed attributes. Saroj Kr. Biswas et al. [24] extract rules from the NN utilizing pedagogical approach for rule extraction. Guido Bologna [25] present a case study on rule extraction from ensembles of NN and SVM. This section will give an overview of two classifiers that have been used in the proposed work. They are J48 and NB classifiers. As we mentioned previously these classifiers are not RBCs, so we need to extract rule-set from both of them to be used as RBCs. The first subsection will show the major approaches for constructing rule-set from classification models. Thereafter, the next subsection will give a simple overview of statistical NB classifier and present a new method to construct rule-set from it while the last subsection will describe J48 classifier and show how to extract rule-set from it.

### 4.1 Rule Construction Approaches

Andrews et al. [26], divide rule construction techniques from NN which can be extended to other models into two major approaches: **Decompositional** and **Pedagogical** (learning-based). The key distinguishing of these approaches is the way in which the rules have been constructed. The former focuses on extracting rules at the level of individual components of the underlying machine learning method. The latter treats the original model as black-box, hence, need to cooperate with another machine learning technique with inherent explanation capability. This imposes to use the original model (black-box) to generate instances for the second model which in turn, generate rules as for output.

### 4.2 Naïve Bayes Classifier

NB is one of the most efficacious classifiers based on Bayesian Theorem. This classifier learns the conditional probability of the attributes in training data, that is, for each attribute Ai of a known class C, the conditional probability is calculated the likelihood of class C given a certain instance of A1, A2…. An, after that the class will be predicted with maximum posterior probability [27].

NB makes strong, independent assumption among the attributes given the value of class C, so all attributes Ai are conditionally independent given class C [28] and this property could be a weak point if the assumption is not met. The relation between attributes and class label of NB could be expressed in simple structures (see Fig. 1).

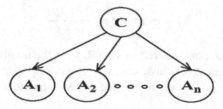

**Fig. 1.** The structure of the Naive Bayes

The Bayes theorem is demonstrated in Eq. 19 below:

$$P\,(C|X) = P(X|C).P(C)/P(X) \tag{19}$$

where, (P) represent probability, $P(C|X)$ is the posterior probability of the class (target) given predictor (attribute), $P(C)$ is a prior probability of a class, $P(X|C)$ denote a conditional probability (Likelihood) of X given class C and $P(X)$ is the prior probability of predictor. The denominator $P(X)$ will be removed when computing $P(C|X)$ since it is constant for all classes as shown in Eq. 20 [5] below:

$$P\,(C|X) \sim P(X|C).p(C) \tag{20}$$

the symbol "~" indicates that the Left Hand Side (LHS) is proportional to the Right Hand Side (RHS). The winning class will be the one that has a maximum posterior probability over the classes [29].

**Naïve Bayes Rule Construction**

The proposed rule construction form NB is considered as a decompositional method. It is possible to derive a set of rule from the NB model. In order to construct rules, the straightforward way is to construct a rule for every attribute of patterns. It could take the easy form as follows:

Rule (R): $X_i ====> Y_k: L$

Where $Xi(X_1, X_2, X_3.... X_n)$ represents the pattern with its attributes, and $Y_k$ ($Y_1$, $Y_2... Y_k$) represent the class label. The last notation (L) represents the power or strength of the rule. This rule construction strategy will produce a large number of rules if the dataset is large. In order to reduce the number of rules, we need a parameter that controls this process (i.e. rule construction), and the rule strength (L) will be that parameter. By utilizing L, it could be eliminating rules with low strength that affect to the hypothesis and hence increase the accuracy of (RBC).

Due to the above observation, the algorithm of the rule construction which is inspired from [30] could be viewed in the figure below (see Fig. 2).

```
Algorithm: rule constructing from NB
Input   : NB conditional probability p(X_i|C),threshold T
Output  : Rule-set (R)
1   Begin
2          for each instance X_i do
3for each value of class C do
4              P:= p(X_i|C_k)
5                if (P>=T)OR (P<=1-T)then
6        R=RU(X_i --> C_k:f(p))
7              end
8            end
9     End
```

**Fig. 2.** An algorithm for constructing rule-set from NB model.

The input of the algorithm is Naïve Bayes conditional probability P(X|C), and threshold T, the output is a set of IF-Then rules. The threshold T will control the number of generated rules. The function f(p) in line (6) represents the label or strength of a rule. It can take one of two forms. The first one is a probabilistic label, that is $f(p) = p$, in this case, if many rules trigger an example and the rules consequents refer to the same class, their consequents should be aggregated via multiplying the labels of rules. Another possibility form is using discretized labels by applying $f(p) = Round(p)$ and the same aggregation procedure is used [30]. In the above algorithm, the condition in line (5) is used when the input domain is restricted to two values which have conditional probability close to 1 or 0. For a domain with more than two values, the condition of line (4) should be replaced with Entropy condition.

Entropy condition for instance $X_i$ for class $Y_k$ can be defined in Eq. 21 as follows:

$$E(i, k) = \sum_{j=1}^{|Dx_i|} -P(\text{Xi} = x_j^i | Y = y_i) log_2 P(\text{Xi} = x_j^i | Y = y_i) \tag{21}$$

where $E(i, k)$ represents entropy of probability distribution, is input domain. The condition in line (4) should have the form: $E < T$.

### 4.3   Decision Tree J48 Classifier

Decision Tree (henceforth, DT) algorithm tries to find out the relationship between the attributes and the patterns in the dataset, in other words, it reveals how the attribute-vector thinking for a number of patterns in the dataset. So based on this concept, DT could obtain the class label for newly unseen patterns [31].

J48 is an expansion of the C4.5 algorithm and was implemented in Weka tool as a DT classifier [32]. Several features are added to J48 classifier, some of them are adapting for missing values, pruning the tree, derivation of rules, etc. J48 was able to generate rule-set from the dataset that can be used for classification tasks [33].

### J48 Rule Construction

Rule extraction from DT is done in a very simple and straightforward way. The rule-set is obtained from every path in the tree starting from the root and ending with the leaf that expresses the class [22].

Due to the above observation, the general algorithm of the rule construction from DT which is inspired from [22] could be viewed in the Fig. 3 as follows:

```
Algorithm: rule extraction from J48
Input   : Training dataset
Output  : Classification rule-set
1 Begin
2           Build J48 model using provided training data.
3           Apply J48 pruning procedure on generated tree.
4           For each path in the tree do
5Extract rule from the root to the leaf
6   end
7   End
```

**Fig. 3.** Algorithm of rule-set extraction from DT (J48)

## 5   Methodology

Our proposed methodology for utilizing rule evaluation metrics as conflict resolution during classification tasks when combining two heterogeneous classifiers could be overviewed in the block diagram below (see Fig. 4).

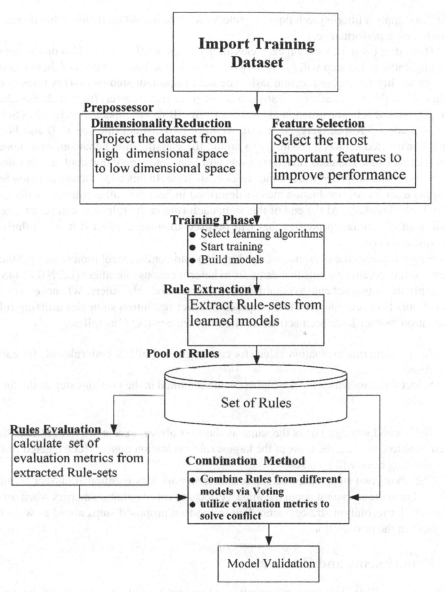

**Fig. 4.** Proposed system architecture

First of all (step 1), we select a dataset for training purpose. For this work, the selected datasets are (WBC, Vote, and Diabetes). The second step (step 2) refers to as a preprocessor which includes some dimensionality reduction and feature selection techniques which are necessary if the selected dataset is huge and contains a large number of instances and/or attributes. Also, dimensionality reduction is essential to reduce the overall dataset size and take a representative sample from it that reflect the behaviour of the original data. This could be done by using special techniques like Sampling. For this work, we apply a filtering technique[1] to select the most important features that increase classification performance.

Thereafter (step 3), we train each classification model (i.e. NB & J48) on the same training dataset, this step will produce two diverse learned models each of them has its own capability for a classification task. The next important step (step 4) is rules construction. In this step, authors construct a set of rules (i.e. rule-set) from each classifier using proposed rule construction algorithms from NB & J48 models described in Subsect. 4.2 and 4.3 respectively. After that, we have two distinct rule-sets (NB and J48). Each constructed set has a capability to mimic the classification behaviour of a model that it came from. The constructed rule-sets will pour into a pool that contains all rules.

Step 5 of the proposed work named rules evaluation. In this step, authors have tended to apply a set for rule evaluation metrics described in Sect. 3 to all constructed rule-sets from both classifiers. At the end of this step, each rule in the rule-sets will be attached with a set of evaluation metrics that will be used to solve conflict if it arises during classification tasks.

Step 6 includes the operation of combination and conflict resolution strategy. Since there exist different rule-sets that came from heterogeneous classifiers (i.e. NB & J48), we apply the voting scheme to combine the heterogeneous classifiers. Whenever a conflict occurs between rule-sets, two proposed conflict resolution strategies utilizing rule evaluation metrics have been activated. The first strategy (A) is as follows:

1) Accumulate rule evaluation values for each triggered rule in both rule-sets for each class.
2) Select the class which has the highest value obtained in the previous step as the final class.

The second strategy (B) is the same as the first above, except we did not take the accumulation, instead, the class of the largest rule evaluation metrics for triggered rules as a winning class will be taken.

The above two steps could be applied for every metrics mentioned in Sect. 3, and this will provide an insight about the impact effects of rule evaluation metrics when used as a conflict resolution strategy according to the two proposed steps above as we will see later in the next section.

## 6    Experiments and Results

This section will explain the experiments that have been conducted on elected datasets. Thereafter, the results obtained by applying the proposed approach will be presented.

### 6.1    Experiments

For this work, three different datasets have been selected: Wisconsin Breast Cancer (WBC), Vote, and Diabetes datasets. These datasets are available at the UCI repository

---

[1] According to Filter Technique, features are selected in isolation from data mining algorithms (i.e. before classification model is run) using some techniques to select the most appropriate features [22].

for machine learning tasks [34]. Vote include nominal attributed while the rest contain numerical attributes. Table 3 shown below summarizes these datasets. WBC contain patient data with two classes representing benign and malign of cancer. In Vote data, the class represents the party of vote records of congressmen. Diabetes dataset is originally from the National Institute of Diabetes and Digestive and Kidney Diseases. It predicts whether or not the patient has diabetes based on several observations and measurements in the data.

**Table 3.** Dataset summary

| Dataset | No. of classes | No. of attributes | Missing value | No. of nominal attribute | No. of numerical attribute | No. of instances |
|---------|----------------|-------------------|---------------|--------------------------|----------------------------|------------------|
| WBC | 2 | 9 | Several | 0 | 9 | 699 |
| Voting | 2 | 16 | Many | 16 | 0 | 435 |
| Diabetes | 2 | 9 | None | 0 | 9 | 768 |

Weka tool was used to build the NB & J48 model for each of the above datasets. This tool contains groups of a machine learning algorithm for data mining tasks. The most attractive feature in Weka is that all algorithms which it supported can either be applied straightway via GUI or called from Java code [35]. Note that all the following experiments were performed using software developed by the authors.

## 6.2 Results

To complete our proposed approach, rule-set from the NB model should be constructed. According to the algorithm presented in Subsect. 4.2, the rule-set has been constructed under the configuration shown in Table 4 below.

**Table 4.** Configuration applied to construct rule-set from NB

| Datasets | Discretization method | No. of bin | Labeling scheme | Threshold T | No. of generated rules |
|----------|----------------------|------------|-----------------|-------------|------------------------|
| WBC | Supervised | 4 | $f(p) = round(p)$ | 0.9 | 12 |
| Vote | | 1 | | | 13 |
| Diabetes | | 3 | | | 9 |

Discretization is an essential technique that has been used by our presented NB rule construction algorithm. It helps to reduce the complexity of the dataset by dividing attributes in the dataset into ranges. In fact, there are two approaches for discretization. The former is supervised, in which the process of discretization will take into account the class label of the record when producing attribute ranges. The latter is unsupervised, that work as blind in the sense of the class of a record when conducting discretization. Our experiments have used the first approach of discretization since it produces relatively high accuracy for classification tasks.

Another factor that affect to the number of a generated rule is the number of the bin using in discretization process. This parameter also has a magical influence on the accuracy of the classification rules and it is selected after extensive experiments on each dataset.

The labeling $f(p) = round(p)$ scheme and threshold $T = 0.9$ have been adopted in this work since these configurations produce a minimum number of rules with relatively high accuracy.

After constructing rule-sets from both classifiers (i.e. NB & J48), the authors have applied a set of evaluation metrics described in Sect. 3 on each constructed rule-set and attached the generated values to each rule. Table 5 will show rule evaluation metrics that attached to rule-sets of Vote dataset.

The results on Diabetes dataset indicate that the empirical metric *cons* show high classification accuracy, followed by *Qprod* compared with other empirical metrics in both strategies (i.e. Strategy A & B) (see Fig. 5), while experiments conducted on a WBC dataset with empirical metrics indicate that all metrics have been given relatively high accuracy in both strategies (see Fig. 6).

The plot in Fig. 7 reflects the accuracy of empirical metrics when applied to Vote dataset. In this dataset, the conflict strategy B shows spurious accuracy compared with strategy A.

The variation in the accuracy of (*Qws* and *Qprod*) in three datasets shown in Figs. 6, 7 and 8 is due to the fact that these metrics depend on consistency and coverage metrics, that is, the degradation of one metric will affect the overall accuracy. One can be made tolerate between these metrics. For example, in *Qws*, we can give more weight to the accuracy rather than coverage by adjusting (w1 and w2).

Table 6 shows the accuracy obtained from statistical evaluation metrics when applied as a conflict resolution strategy in Diabetes dataset.

From the above table (Table 6), it seems that X2 shows degradation in accuracy in both strategies, also $Q_{cohen}$ metric has low accuracy compared with another measure of agreement metrics. On the other hand, the measure of logical sufficiency shows high and comparable accuracy in both strategies (i.e. A & B). Finally, Laplace and M-estimate metrics gave superior accuracy in strategy B compared with all statistical metrics in Diabetes dataset.

The plot shown in Fig. 8 expresses the accuracy of statistical evaluation metrics when utilized as conflict resolution on WBC dataset. It is clear that the overall accuracy of statistical metrics shows high accuracy that reaches 97.4% in $Q_{IS}$ metric, but there area few degradations in accuracy with the measure of agreement especially in strategyB.

The sketch in Fig. 9 shows the comparison of rule evaluation metrics on Vote dataset, our observation in this figure shows that the B strategy produces better accuracy than A strategy except on three statistical metrics named ($Q_{coleman}$, $Q_{c1}$, and $Q_{c2}$).

**Table 5.** Rule evaluation metrics for Vote dataset

| No | cons | cover | QWS | Prod | X2 | G2 | CChen | Coleman | QC1 | QC2 | QIS | QLS | QMD | Laplace | M-estimate |
|---|---|---|---|---|---|---|---|---|---|---|---|---|---|---|---|
| NB R0 | 1.0 | 0.9 | 1.0 | 0.9 | 344.7 | 0.9 | 409.9 | 1115.6 | 153189.5 | 1069.7 | 0.7 | 77.1 | 9.9 | 1.0 | 1.0 |
| NB R1 | 1.0 | 0.7 | 0.9 | 0.7 | 203.3 | 0.7 | 359.0 | 1081.4 | 130139.1 | 945.7 | 0.6 | 15.7 | 5.9 | 1.0 | 1.0 |
| NB R2 | 0.9 | 0.5 | 0.8 | 0.5 | 74.2 | 0.4 | 285.1 | 998.8 | 95592.5 | 751.9 | 0.5 | 5.0 | 3.2 | 0.9 | 0.9 |
| NB R3 | 0.7 | 0.5 | 0.6 | 0.4 | 3.1 | 0.0 | 233.0 | 736.9 | 57727.2 | 560.3 | 0.1 | 1.2 | 0.5 | 0.7 | 0.7 |
| NB R4 | 0.9 | 0.8 | 0.9 | 0.7 | 191.0 | 0.5 | 360.0 | 1028.1 | 124042.3 | 924.1 | 0.6 | 6.7 | 4.9 | 0.9 | 0.9 |
| NB R5 | 0.9 | 0.7 | 0.8 | 0.6 | 120.7 | 0.4 | 324.0 | 1001.5 | 108837.6 | 836.4 | 0.5 | 5.1 | 3.8 | 0.9 | 0.9 |
| NB R6 | 1.0 | 0.6 | 0.9 | 0.7 | 159.9 | 0.8 | 331.1 | 1104.9 | 122675.8 | 898.0 | 0.7 | 35.0 | 6.5 | 1.0 | 1.0 |
| NB R7 | 0.6 | 0.8 | 0.7 | 0.5 | 71.8 | 0.1 | 298.0 | 401.8 | 40182.4 | 361.1 | 0.6 | 2.1 | 2.7 | 0.6 | 0.6 |
| NB R8 | 0.8 | 0.8 | 0.8 | 0.7 | 234.5 | 0.8 | 378.9 | 587.9 | 74641.8 | 542.4 | 1.1 | 7.8 | 5.5 | 0.8 | 0.8 |
| NB R9 | 0.7 | 0.7 | 0.7 | 0.5 | 110.7 | 0.3 | 329.9 | 478.3 | 52923.2 | 414.3 | 0.8 | 3.3 | 3.3 | 0.7 | 0.7 |
| NB R10 | 0.7 | 0.8 | 0.8 | 0.6 | 165.6 | 0.5 | 353.9 | 528.9 | 62748.9 | 473.8 | 1.0 | 4.7 | 4.2 | 0.7 | 0.7 |
| NB R11 | 0.7 | 0.9 | 0.8 | 0.6 | 171.7 | 0.4 | 352.0 | 501.7 | 59191.7 | 468.8 | 0.9 | 3.9 | 4.5 | 0.7 | 0.7 |
| NB R12 | 0.5 | 0.8 | 0.6 | 0.4 | 52.8 | 0.1 | 278.1 | 369.8 | 34529.1 | 336.8 | 0.4 | 1.7 | 2.4 | 0.5 | 0.5 |
| NB R13 | 0.6 | 0.8 | 0.7 | 0.5 | 105.5 | 0.2 | 317.0 | 431.3 | 45864.8 | 397.9 | 0.7 | 2.5 | 3.4 | 0.6 | 0.6 |
| J48 R0 | 1.0 | 0.8 | 0.9 | 0.8 | 298.9 | 1.7 | 397.8 | 692.7 | 92314.4 | 624.6 | 1.3 | 71.5 | 8.5 | 1.0 | 1.0 |
| J48 R1 | 0.9 | 0.1 | 0.6 | 0.4 | 20.6 | 1.0 | 280.5 | 606.8 | 57140.0 | 335.9 | 1.2 | 9.5 | 3.4 | 0.8 | 0.8 |
| J48 R2 | 1.0 | 0.0 | 0.8 | 0.4 | 3.2 | 0.0 | 267.4 | 708.1 | 63589.6 | 358.3 | 1.4 | 0.0 | 0.0 | 0.8 | 0.7 |
| J48 R3 | 1.0 | 0.9 | 1.0 | 0.9 | 344.7 | 0.9 | 409.9 | 1115.6 | 153189.5 | 1069.7 | 0.7 | 77.1 | 9.9 | 1.0 | 1.0 |
| J48 R4 | 1.0 | 0.0 | 0.8 | 0.4 | 2.5 | 0.0 | 171.4 | 1124.8 | 64997.5 | 570.8 | 0.7 | 0.0 | 0.0 | 0.8 | 0.9 |
| J48 R5 | 0.8 | 0.0 | 0.6 | 0.3 | 0.7 | 0.2 | 170.4 | 899.5 | 51679.4 | 456.5 | 0.4 | 2.5 | 1.3 | 0.7 | 0.7 |

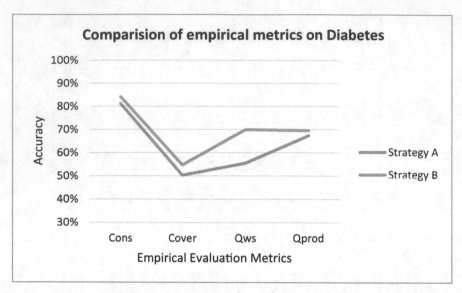

**Fig. 5.** The impact of empirical evaluation metrics on Diabetes dataset

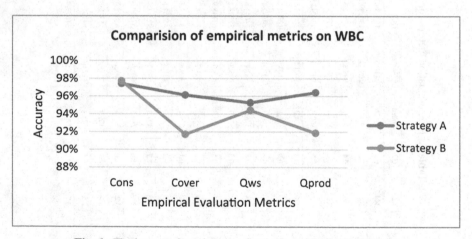

**Fig. 6.** The impact of empirical evaluation metrics on WBC dataset

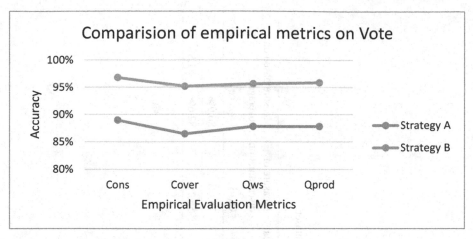

**Fig. 7.** The impact of empirical evaluation metrics on Vote dataset

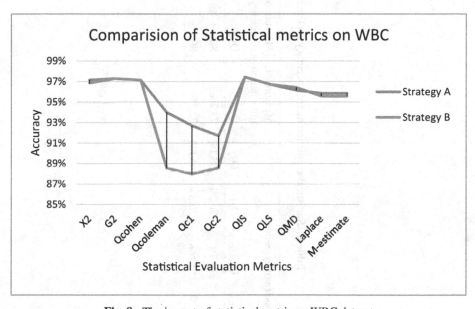

**Fig. 8.** The impact of statistical metric on WBC dataset

**Table 6.** Statistical evaluation metrics accuracy on Diabetes dataset

| Strategy | Measure of association | | Measure of agreement | | | | Measure of information | Measure of logical sufficiency | Measure of discrimination | Measure based on coverage | |
|---|---|---|---|---|---|---|---|---|---|---|---|
| | $X^2$ | G2 | $Q_{cohen}$ | $Q_{coleman}$ | $Q_{c1}$ | $Q_{c2}$ | $Q_{IS}$ | $Q_{LS}$ | $Q_{MD}$ | Laplace | M-estimate |
| A | 28.8% | 76.2% | 37.2% | 74.7% | 76.2% | 68.1% | 73.7% | 80.1% | 75.0% | 80.9% | 81.0% |
| B | 34.4% | 78.0% | 74.7% | 65.6% | 70.3% | 65.6% | 72.7% | 80.1% | 55.7% | 82.6% | 81.8% |

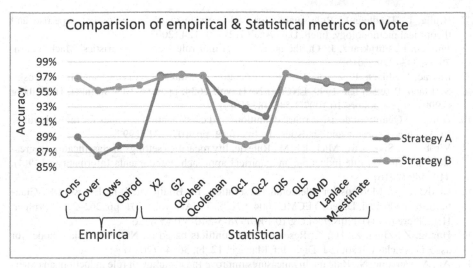

**Fig. 9.** The impact of rule evaluation metrics on Vote dataset

## 7  Conclusion

This paper presents a description of empirical and statistical rule evaluation metrics. Also, the work presents an approach that utilizes and shows the impact of these metrics as a rule conflict resolution strategy during classification tasks when combining two heterogeneous classifiers (NB and J48) utilizing two suggested conflict resolution strategies (A & B).

The experimental results indicate that strategy B shows overall better accuracy compared with strategy A. The most important conclusion of this paper is that the impact of both empirical and statistical evaluation metrics when used as conflict resolution strategies differ from one dataset to another for the same metric. This means that datasets behaviour also has a different impact on conflict resolution in specific metrics. Moreover, our observation to statistical metrics shows that *Cons*, *Laplace*, and *M-estimate* metrics showed superiority with respect to accuracy compared with other metrics.

## References

1. Michalski, R.S.: Pattern recognition as rule-guided inductive inference. IEEE Trans. Pattern Anal. Mach. Intell. **4**, 349–361 (1980)
2. Clark, P., Niblett, T.: The CN2 induction algorithm. Mach. Learn. **3**(4), 261–283 (1989)
3. An, A., Cercone, N.: Rule quality measures for rule induction systems: description and evaluation. Comput. Intell. **17**(3), 409–424 (2001)
4. Bruha, I., Kockova, S.: Quality of decision rules: empirical and statistical approaches. Informatica **17**, 233–243 (1993)
5. Han, J., Pei, J., Kamber, M.: Data Mining: Concepts and Techniques. Elsevier, Amsterdam (2011)
6. Webb, A.R., Copsey, K.D.: Statistical Pattern Recognition. Wiley, Hoboken (2011)

7. Bruha, I., Tkadlec, J.: Rule quality for multiple-rule classifier: empirical expertise and theoretical methodology. Intell. Data Anal. **7**(2), 99–124 (2003)
8. Janssen, F., Fürnkranz, J.: On the quest for optimal rule learning heuristics. Mach. Learn. **78**(3), 343–379 (2010)
9. Lavrač, N., Flach, P., Zupan, B.: Rule evaluation measures: a unifying view. In: Džeroski, S., Flach, P. (eds.) ILP 1999. LNCS (LNAI), vol. 1634, pp. 174–185. Springer, Heidelberg (1999). https://doi.org/10.1007/3-540-48751-4_17
10. Bruha, I.: Quality of decision rules: definitions and classification schemes for multiple rules. In: Machine Learning and Statistics, the Interface, pp. 107–131 (1997)
11. Wróbel, Ł., Sikora, M., Michalak, M.: Rule quality measures settings in classification, regression and survival rule induction: an empirical approach. Fundamenta Informaticae **149**(4), 419–449 (2016)
12. Lindgren, T.: Methods for rule conflict resolution. In: Boulicaut, J.-F., Esposito, F., Giannotti, F., Pedreschi, D. (eds.) ECML 2004. LNCS (LNAI), vol. 3201, pp. 262–273. Springer, Heidelberg (2004). https://doi.org/10.1007/978-3-540-30115-8_26
13. Huang, Z., Zhou, Z., He, T.: Resolving rule conflicts based on Naïve Bayesian model for associative classification. J. Digit. Inf. Manage. **12**(1), 36–43 (2014)
14. An, A., Cercone, N.: Rule quality measures improve the accuracy of rule induction: an experimental approach. In: Raś, Zbigniew W., Ohsuga, S. (eds.) ISMIS 2000. LNCS (LNAI), vol. 1932, pp. 119–129. Springer, Heidelberg (2000). https://doi.org/10.1007/3-540-39963-1_13
15. Bishop, Y.M., Fienberg, S.E., Holland, P.W.: Discrete Multivariate Analysis: Theory and Practice. The MIT Press, Cambridge (1977)
16. Han, J., Kamber, M., Pei, J.: Data Mining Concepts and Techniques, 3rd edn. Morgan Kaufmann, Burlington (2011)
17. Torgo, L.: Controlled redundancy in incremental rule learning. In: Brazdil, P.B. (ed.) ECML 1993. LNCS, vol. 667, pp. 185–195. Springer, Heidelberg (1993). https://doi.org/10.1007/3-540-56602-3_136
18. Torgo, L.: Knowledge integration. Curr. Trends Knowl. Acquis. **8**, 90 (1990)
19. Cohen, J.: A coefficient of agreement for nominal scales. Educ. Psychol. Measur. **20**(1), 37–46 (1960)
20. Kononenko, I., Bratko, I.: Information-based evaluation criterion for classifier's performance. Mach. Learn. **6**(1), 67–80 (1991)
21. An, A., Cercone, N.: ELEM2: a learning system for more accurate classifications. In: Mercer, R.E., Neufeld, E. (eds.) AI 1998. LNCS, vol. 1418, pp. 426–441. Springer, Heidelberg (1998). https://doi.org/10.1007/3-540-64575-6_68
22. Tan, P.-N., Steinbach, M., Kumar, V.: Introduction to Data Mining. Pearson, London (2005)
23. Chakraborty, M., Biswas, S.K., Purkayastha, B.: Recursive rule extraction from NN using reverse engineering technique. New Gener. Comput. **36**(2), 119–142 (2018)
24. Biswas, S.K., Chakraborty, M., Purkayastha, B.: A rule generation algorithm from neural network using classified and misclassified data. Int. J. Bio-Inspired Comput. **11**(1), 60–70 (2018)
25. Bologna, G., Hayashi, Y.: A rule extraction study from SVM on sentiment analysis. Big Data Cogn. Comput. **2**(1), 6 (2018)
26. Andrews, R., Diederich, J., Tickle, A.B.: Survey and critique of techniques for extracting rules from trained artificial neural networks. Knowl.-Based Syst. **8**(6), 373–389 (1995)
27. Friedman, N., Geiger, D., Goldszmidt, M.: Bayesian network classifiers. Mach. Learn. **29**(2–3), 131–163 (1997)
28. Možina, M., Demšar, J., Kattan, M., Zupan, B.: Nomograms for visualization of naive Bayesian classifier. In: Boulicaut, J.-F., Esposito, F., Giannotti, F., Pedreschi, D. (eds.) PKDD 2004. LNCS (LNAI), vol. 3202, pp. 337–348. Springer, Heidelberg (2004). https://doi.org/10.1007/978-3-540-30116-5_32

29. Alashqur, A.: A novel methodology for constructing rule-based naïve Bayesian classifiers. Int. J. Comput. Sci. Inf. Technol. **7**(1), 139–151 (2015). https://doi.org/10.5121/ijcsit.2015. 7114

30. Śnieżyński, B.: Converting a Naive Bayes model into a set of rules. In: Kłopotek, M.A., Wierzchoń, S.T., Trojanowski, K. (eds.) Intelligent Information Processing and Web Mining, pp. 221–229. Springer, Heidelberg (2006). https://doi.org/10.1007/3-540-33521-8_22

31. Korting, T.S.: C4. 5 algorithm and multivariate decision trees, image processing division. National Institute for Space Research–INPE, SP, Brazil (2006)

32. Hall, M., Frank, E., Holmes, G., Pfahringer, B., Reutemann, P., Witten, I.H.: The WEKA data mining software: an update. ACM SIGKDD Explor. Newslett. **11**(1), 10–18 (2009)

33. Kaur, G., Chhabra, A.: Improved J48 classification algorithm for the prediction of diabetes. Int. J. Comput. Appl. **98**(22), 13–17 (2014)

34. Blake, C.L., Merz, C.J.: UCI Repository of Machine Learning Databases. University of California, Irvine (1998)

35. Weka 3: Data mining software in Java. University of Waikato, Hamilton, New Zealand, 19 52 (2011). www.cs.waikato.ac.nz/ml/weka

# Parallel Genetic Algorithm for Optimizing Compiler Sequences Ordering

Manal H. Almohammed, Ahmed B. M. Fanfakh$^{(\boxtimes)}$, and Esraa H. Alwan

Department of Computer Science, College of Science for Women, University of Babylon, Hillah, Iraq
manalalmohamed.h@gmail.com,
{ahmed.badri,wsci.israa.hadi}@uobabylon.edu.iq

**Abstract.** Recent compilers provided many of optimization passes, thus even for expert programmers it is really hard to know which compiler optimization among several optimizations passes can improve program performance. Moreover, the search space for optimization sequences is very huge. Therefore, finding an optimal sequence for even a simple code is not an easy task. The goal of this paper is to find a set of a good optimization sequences using parallel genetic algorithm. The method firstly classifies the programs into three clusters then applying three versions of genetic algorithms each one to cluster in parallel. In order to enhance the result, the migration strategy between these three algorithms is applied. Three optimal sequences at the same time are obtained from this method. However, the proposed method improved the execution time on average by 87% compared with the O2 optimization flag. This method also outperforms the sequential version of genetic algorithm on average of the execution time by **74.8%** in case of using Tournament selection and **72.5%** in case of multi-selection method. LLVM framework is used to validate and execute the proposed method. In addition, Polybench, Standerford, Shootout benchmarks are used as case study to verify the effectiveness of the proposed method.

**Keywords:** Phase ordering · Compiler optimization · Parallel genetic algorithm · Performance

## 1 Introduction

Recentcompilers introduce a large number of optimizations such as elimination dead code, replacing certain methods/functions with superior versions, reordering code block, numbering global value, factor, etc. In particular, the order of applying these optimizations can affects the form of code for subsequent optimization phases. However, any optimization applied to the code may improve one or more of the performance metrics such as power, code size, and execution time. Some times when using these optimizations for the program, these optimizations interact with each other in a complex way. This interaction can affect program performance or may produce bad or incorrect code. This problem is known as the phase ordering problem [1]. Therefore, standard compilers

© Springer Nature Switzerland AG 2020
A. M. Al-Bakry et al. (Eds.): NTICT 2020, CCIS 1183, pp. 128–138, 2020.
https://doi.org/10.1007/978-3-030-55340-1_9

introduced optimization levels that are applied for a fixed order, or in some pre-set order that seems to produce comparatively good results. However, many authors have shown that the performance of the application depends mainly on the choice of optimizations by selecting the best order. In the literature, each group of passes with certain order called sequence of passes. Therefore, selecting automatically the best ordering of compiler optimization for any program is difficult task and has significant importance in field of the compiler for a long time [2].

Many researchers have been proposed heuristics and meta-heuristics methods to solve the problem of passes ordering see [3] and [4]. To solve this problem, we proposed in this paper a method to find global optimization sequences of passes and be comprehensive for all programs that give free of interaction and dependency. In our previous work [5], the proposed method uses performance counters to classify the benchmark to three classes. Then, genetic algorithm was executed on each class sequentially. Where each algorithm works independently from each other. However, in this work parallel genetic algorithms are proposed by applying migration strategy to improve the obtained results. This paper is organized as follows: the next Sect. (2) describes how to extract performance events from the performance counter. Section (3) describes the proposed method where the implementation of parallel Genetic Algorithm (PGA) has been explained. Section (4) presents the experimental results obtained from the application of the proposed method. Finally, in Sect. (5), some of the related works have been introduced and the last section, Sect. (6), displays the conclusions and future work.

## 2  Related Work

T. Satish Kumar et al. present parallel Genetic algorithm that selected the compiler flags to optimize code for multicore CPUs. To compare and test the performance, three methods had been adopted. The first methods compiled the benchmark without applying optimization. The second used randomly selected optimization flags and finally, the compiler optimization levels had been set using parallel genetic algorithm [7]. Pan and Eigenmann [8] introduce a new algorithm called Combined Elimination (CE) which is able to tune the program performance by picking the optimizations passes that can improve the program performance. It can get better or comparable performance with less time to tune the optimization sequence. Kulk arni et al. [9] proposed a comprehensive search strategy to find the best optimization sequences for each of the functions in a program. The relationship between different phases is calculated automatically by doing a comprehensive analytic to these phases. According to these analytical results the execution time is reduced. Cooper, Schielke, and Subramanian [10] used a method to minimize the code size. They used a genetic_algorithm_based approach to find best sequences. Triantafyllis et al. [11] proposed an iterative compilation which is applying various optimization configuration on the same program code segment. Then a comparison among different versions of optimized code segment has occurred and the best one will be chosen. In [12] two complementary general methods described to reduce the search time for finding the optimization sequence using genetic algorithm (GA). The search time is reduced in the first method by avoiding the unnecessary execution of the application. While in second method the same results achieved by modifying the search using a fewer number of generations. The proposed method mainly different from other methods by applying the migration strategy synchronously in parallel.

## 3  Feature Extraction Using Performance Counter

The information extracted from the performance counter called performance event, which represents the dynamic behavior of the programs. These events consider an important aspect of program improvement, for example, instructions, cache_references, emulation_faults, and others [6] and [7]. On the other hand, these events are an event-oriented portability tool, which can be used to help in solving forward improvement and troubleshooting service [2]. The use of performance counters is attractive because they do not limit the program class, which the system is able to handle. As a result, our system (strategy) can find a good compiler optimization sequence for any program.

## 4  Proposed Method

In this paper, parallel genetic algorithms are proposed to work on each cluster. Each one used different genetic selection method depending on the offline decision and experiments. Each selection method decided to each cluster by running the genetic algorithm three times one for each cluster. However, for the three different clusters there are three different versions of genetic algorithms that used different selection methods. Moreover, these three genetic algorithms are executed synchronously in parallel, each one executed over different core of the CPU. Thus, multi-core architecture of the CPU is used to run the three algorithms in parallel using message passing interface (MPI) under C language. MPI is the widest parallel library that works under C or Fortran [13, 14]. Migration strategy is applied to all parallel system by sending the best solutions between the algorithms after some iterations. This strategy used to improve the results by increasing the search space diversity. The resultant of the best sequence for each cluster can be considered as an optimization sequence for any unseen program similar to the features of that cluster.

Table 1 presents the features used in our approach. The first column lists the perf events; the second column gives the used type.

The following is the outline of the proposed method:

1-Extracting performance events by using compiler -O0.
2-Computed the similarity for each program used in this method as follows:

$$\text{Sim}(p, pi) = \frac{\sum_{w=1}^{m} Pw \times Piw}{\sqrt{\sum_{w=1}^{m} (Pw)^2}\sqrt{\sum_{w=1}^{m} (Piw)^2}}$$

where $p$ and $pi$ represent the base program and other programs respectively [15].
3-The programs are divided into three clusters according to them similarity results.
4-Running the three genetic algorithms over each cluster synchronously in parallel.
5-Iteratively, migration is applied after a predefined number of generations between the algorithms during parallel execution.

**Table 1.** The used features in the proposed approach

| Event | Type |
|---|---|
| Cpu-cycles or cycles, instructions, Cashe- references, Cashe-misses, Bus_cycles | Hardware event |
| Cpu_clock(msec), Task_clock(msec), Page- faults OR faults, Context-switches, Cpumigrations, Alignment-faults, Emulationfaults | Software event |
| L1-dcashe-loads, L1-dcashe-loads misses, L1-dcashe-stores, L1-dcashe-storesmisses, L1-icashe-loads, L1-icashe-loadsmisses, L1-icashe-loads, L1-icashe-loads misses, L1-icashe- prefetches, L1-icasheprefetches –misses, LLC-load, LLC-loadsmisses, LLC-strores, LLC-strores misses, LLC-prefetch-misses, Dtlb-loads, Dtlb-loadsmisses, Dtlb-store, Dtlb-store-misses, Dtlb prefetches, Dtlb-prefetches -misses, Itlb -loads, Itlb –loads -misses, Branch-loads, Branch-loads-misses | Hardware cache event |
| Sched:sched-stat-runtime, Sched:sched-pi-setprio, Syscalls:sys_enter_socket, Kvmmmu:kvm_mmu_pagetable_walk | Tracepoint event |
| Stalled_cycles-frontend, Stalled_cycle- backend | Dynamic event |
| Sched:sched_process_exec, Sched:sched_process_frok, Sched:sched_process_wait,Sched:sched_process_wait_task, Sched:sched_process_exit | Tracepoint event |

6-Determining the best sequence for each cluster after finishing the execution of all genetic algorithms.

7-For any new unseen program, its similarity is computed by extracting its features and comparing them with the features of all clusters.

8-Then, the new unseen program can be matched with the most similar cluster and the optimization sequence of that cluster can be used to optimize the unseen program. Figure 1 illustrate the structure of the proposed method.

More details about the proposed method are presented as follows:

**Step 1: Extracting the Features of the Program**

In this method, 60 programs are used where each program collected 52 events. For getting more accuracy, each program executed three times and the average is computed for 52 events. Therefore, the similarity of each program's features with the features of other programs is computed. Depending on the resultant similarity, clustering methods can be applied, for more details about this step see [5].

**Step 2: Parallel Genetic Algorithm for Sequence Optimization (PGA)**

Genetic algorithm is a metaheuristic search method used to optimize the candidate solutions based on the Darwin evolution principles [6] and [7].In this work, the genetic algorithm is applied to the problem of optimizing sequence ordering. The diversity of the solutions in the genetic algorithm depends mainly on many aspects such as selection,

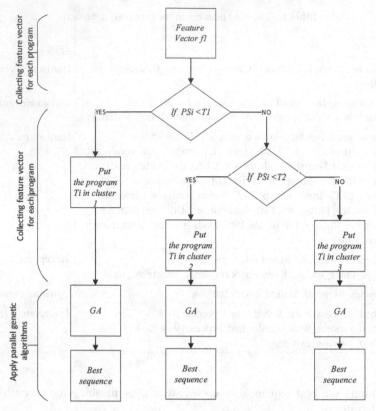

**Fig. 1.** The structure of the proposed method

crossover, mutation methods, and them rates used. One of the most important factors to increase the diversity of the solution in the search space of the genetic algorithm is the migration strategy. Its main idea has subdivided the population of the algorithm into multiple populations and applying migration of the best solution of each population with other ones. The major drawback of this strategy is the high computational cost. To tackle this problem and increasing the diversity of the solutions, three genetic algorithms applying migration strategy are applied synchronously in parallel. Each algorithm, works on specific cluster using different selection operator. Offline experiments are executed to decides the best selection method for each cluster depending on its feature. The selection method in the first algorithm was stochastic universal while in the second algorithm used Rowlett wheel and tournament selection used in the third algorithm. Algorithm (1) is describe the general framework of the three algorithms.

The algorithm used integer representation for the generated chromosomes by maximum length of 64 genes. Each gene can represent a number that may match a pass. To generate variable length chromosomes, the zero-integer value is used to represent the no-pass state. Whereas, the others genes can take an integer number from 1 to 64 which corresponding a specific pass.

**Algorithm (1)**:  Parallel Genetic algorithm for sequence optimization
***Require:***
*Pop_size: The population size of the algorithm in cluster.*

*Opt_passes: The vector of the optimization passes.*
*Chrom_length: The length of the chromosome in the algorithm.*
*Pn:Number of the program in the cluster.*
*Iterations: is the iteration index of the algorithm.*
*Maxgen: is the maximum number of iterations.*
*Fit: Fitness value.*
**Output:** *best optimization sequence of passes for selected cluster of programs.*
*1:  Generating the initial population.*
*3:  **for** i=1 to pop_size do*
*4:     **for** j=1 to chrom_length do*
*5:          pop [i].sequence[j]⇒select randomly a pass from vector Opt_passes*
*6:          Flip ⇒select random number from 0 to 8*
*7:          If flip=1 then*
*8:             pop [i].sequence[j]=0*
*9:       else*
*10:          Pop [i].sequence[j]=select random pass from vector Opt_passes.*
*11:   **endif**.*
*12:   **end for**.*
*13:   Compute the average of execution time for the sequence Pop [i]. sequence[j]*
*         during three runs applied over Pn programs.*
*14:   Pop1[i].fit = average of the execution time for the three runs.*
***15: end for.***
*16: repeat*
*        Iterations =Iterations +1*
*17:   Selecting a chromosome using a one of the selection methods.*
*18:   Applying the uniform crossover over the selected parents.*
*19:   Applying one-point mutation over the two new offspring.*
*20:   Evaluating the new offspring using steps 13 and 14.*
*21:   Replace the best individual from the new produces offspring*
*         instead of the worse individual of a randomly selected group.*
*22: **Until** the standard deviation between the last two generations*
*      reaching to the less fitness function **or Iterations**>maxgen.*

### Step 3: Application of the Migration Between the Three Algorithms

Three parallel genetic algorithms are executed synchronously in parallel over real four-core processor. Three cores are selected to run the three algorithms in parallel using message passing primitives. Moreover, point-to-point communication functionsare used to apply the migration model. Figure 2 shows the migration model between the three parallel algorithms.

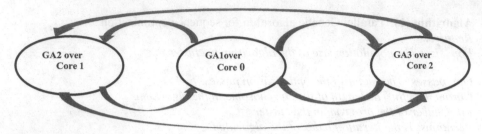

**Fig. 2.** The migration model

MPI_Send() and MPI_Recv communication primitives of MPI are used to migrate the selected individuals between the algorithms during the execution time of the parallel system. After a specific number of iterations, migration is applied between the algorithms.

At each algorithm, three individuals are selected randomly and the best one is the migrated individual. Each received individual from the migration is replaced with worst individual selected from a random group of five individuals if its fitness is better from the later individual.

## 5  Experiments

### 5.1  The Experimental Setup

This work uses LLVM compiler infrastructure and Linux perf tool to validate the proposed method. Message passing library MPI v3.0.2 has been used to program and execute the parallel algorithms. The LLVM Clang (c language frontend) is used to transform the c source code of each program into IR code which is saved in bitcode machine-readable file format (.bc). O2 optimization level is used to measure the effectiveness of the resulted sequences. Next, following in this section a discusses of the results that is obtained from applying the proposed method to programs selected from three benchmarks and comperes it with the previous work [5].

The proposed method uses a collection of 64 LLVM optimizations passes to find a sequence that will give the best or close to optimal execution time for each cluster. In the proposed method, three populations have been initialized, each one uses a population size equal to 100, the probability of crossover is 0.5, the mutation is 0.01, the ratio of occurrence gene 0.4, the standard deviation for the stopping criteria is 0.01. Moreover, the migration is applied after 10 generations. The maximum chromosome length used in all algorithms is equal to 64 genes.

## 5.2  Experimental Results

This section presents the results of the proposed parallel genetic algorithms for optimizing the complier sequence ordering problem. To verify the accuracy and the achievement of the proposed parallel framework, two genetic algorithms that used two different strategies are compared with the proposed method. The first algorithm is used tournament selection, uniform crossover and one-point mutation for all clusters. The algorithm executed three times to obtained the results, we refer to this algorithm TSGA. The second algorithm different from the first one only in the selection method used. This algorithm used different selection methods, where each one is selected to fit the feature of each cluster. The Stochastic universal sampling, Roulette wheel, and Tournament selections methods are used for clusters 1,2,3 respectively. Thus, we refer to this algorithm as MSGA. Figures (3, 4, 5) show the comparison results of the proposed parallel genetic algorithm (PGA) with TSGA and MSGA. Figures also presents the comparison of the three algorithms with the standard optimization flag -O2.

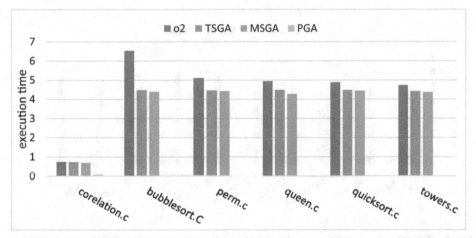

**Fig. 3.** Comparison of the execution time between TSGA, MSGA, PGA and the O2 in the first cluster.

The obtained results show that the proposed method gives better results, less execution time, compared to all other methods. This due to the high diversity that introduced from applying the proposed migration strategy in parallel. Moreover, the proposed parallel version reduces the execution time compared to the MSGA algorithm by 3.7. This ratio is the speedup factor that computed as the ration of the sequential execution time divided by the parallel one.

The average of the execution time for all the four methods are computed and presents in the Table 2. The best optimization sequence resulted from each algorithm is presented in Table 3.

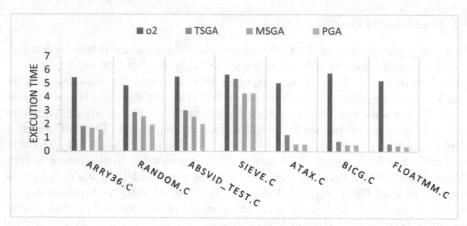

**Fig. 4.** Comparison of the execution time between TSGA, MSGA, PGA and the O2 in the second cluster.

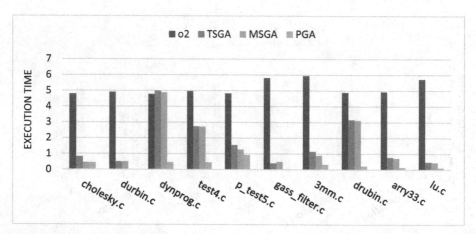

**Fig. 5.** Illustrates the comparison of the execution time between TSGA, MSGA, PGA, and the O2 in the third cluster.

**Table 2.** The comparison of the execution time between O2,TSGA, MSGA and PGA

| Cluster name | O2 | TSGA | MSGA | PGA |
|---|---|---|---|---|
| Cluster 1 | 4.49 | 3.84 | 3.77 | 0.02 |
| Cluster 2 | 5.35 | 2.23 | 1.78 | 1.60 |
| Cluster 3 | 5.15 | 1.66 | 1.55 | 0.32 |

**Table 3.** Best optimization sequence of the second scenario for each cluster

| Cluster No. | Best optimization sequence |
| --- | --- |
| Cluster 1 | -domtree -loop-reduce -mergereturn -sink -gvn -lcssa -loop-simplify -licm -loop-rotate -gvn -instcombine -lcssa -indvars -lowerinvoke -simplifycfg -loop-unroll -indvars -lazy-value-info -dse -loop-instsimplify -targetlibinfo -basiccg -memdep -sccp -globalopt -indvars -simplifycfg -ipconstprop -gvn -scalar-evolution -argpromotion -loop-unswitch -lcssa -early-cse -domtree -lowerswitch -loop-deletion -tailcallelim -early-cse -lower-expect -indvars -loop-rotate -gvn -partial-inliner -reassociate -loop-unroll -lowerswitch -dse -loop-simplify -loops -inline -constprop -loop-simplify -loop-instsimplify -strip-dead-prototypes -globaldce |
| Cluster 2 | -inline -lowerinvoke -scalar-evolution -tailcallelim -gvn -loop-rotate -indvars -codegenprepare -inline -mergereturn -ipconstprop -codegenprepare -mergereturn -targetlibinfo -loop-instsimplify -argpromotion -instcombine -constmerge -mergefunc -lowerinvoke -lower-expect -constprop -loop-unswitch -licm -instsimplify -tailcallelim -partial-inliner -constprop -lcssa -reassociate -basiccg -lowerswitch -basicaa -loop-instsimplify -instsimplify -simplifycfg -lcssa -mergereturn -scalar-evolution -inline -deadargelim -loop-simplify -strip-dead-prototypes -codegenprepare -loops -instcombine -constprop -dse -lowerswitch -lowerinvoke -simplifycfg |
| Cluster 3 | -loweratomic -constmerge -gvn -inline -loop-idiom -globaldce -memcpyopt -constmerge -dce -deadargelim -lower-expect -indvars -gvn -instcombine -memdep -gvn -ipsccp -prune-eh -lowerswitch -licm -ipconstprop -loop-idiom -early-cse -reassociate -constmerge -domtree -simplifycfg -inline -lazy-value-info -inline -memcpyopt -instsimplify -deadargelim -argpromotion -strip-dead-prototypes -tbaa -basiccg -simplifycfg -functionattrs -tbaa -lcssa -loop-simplify -mergereturn -loop-instsimplify -argpromotion -codegenprepare -memcpyopt -loop-deletion -loop-reduce -early-cse -adce -correlated-propagation -simplifycfg |

# 6 Conclusion

This paper described using parallel genetic algorithm to discover the best optimization sequence. Each genetic algorithm runs over one core. Migration was applied between the populations. The proposed method gave three optimal sequences at the same time. Moreover, the proposed method compared with two different sequential versions of genetic algorithm. The comparison results showed that the proposed parallel version outperforms the other two version of the genetic algorithm due to migration strategy used in parallel.

LLVM infrastructure has been used to validate the proposed method. The experiment results obtained indicate the effectiveness of the proposed method when it compared with the earlier work [5]. Moreover, each obtained sequence for a cluster of programs can be used as a guided sequence for the new unseen program. The similarity was computed between the features of the unseen program with features of the each cluster. Therefor the sequence of the most similar cluster is used to optimize the unseen program. In the future, more programs can be used to expand the number of clusters. Thus, the accuracy of the computed similarity between programs' clusters and the unseen program can be increased. Moreover, the parallel implementation of the genetic algorithm will expand to consider the other genetic operators such as crossover and mutation.

# References

1. Sher, G., Martin, K., Dechev, D.: Preliminary results for neuroevolutionary optimization phase order generation for static compilation. In: Proceedings of the 11th Workshop on Optimizations for DSP and Embedded Systems, pp. 33–40 (2014)
2. Alkaaby, Z.S., Alwan, E.H., Fanfakh, A.B.M.: Finding a good global sequence using multi-level genetic algorithm. J. Eng. Appl. Sci. 9777–9783 (2018)
3. Crainic, T.G., Toulouse, M.: Parallel strategies for meta-heuristics. In: Glover, F., Kochenberger, G.A. (eds.) Handbook of Metaheuristics. International Series in Operations Research & Management Science, vol. 57, pp. 475–513. Springer, Boston (2003). https://doi.org/10.1007/0-306-48056-5_17
4. Hertz, A., Widmer, M.: Guidelines for the use of meta-heuristics in combinatorial optimization. Eur. J. Oper. Res. **151**, 247–252 (2003)
5. Almohammed, M.H., Alwan, E.H., Fanfakh, A.B.M.: Programs features clustering to find optimization sequence using genetic algorithm. In: Jain, L.C., Peng, S.-L., Alhadidi, B., Pal, S. (eds.) ICICCT 2019. LAIS, vol. 9, pp. 40–50. Springer, Cham (2020). https://doi.org/10.1007/978-3-030-38501-9_4
6. Nisbet, A.P.: GAPS: Iterative feedback directed parallelization using genetic algorithms. In: Workshop on Profile and Feedback-Directed Compilation (1998)
7. Kumar, T.S., Sakthivel, S., Kumar, S.: Optimizing code by selecting compiler flags using parallel genetic algorithm on multicore CPUs. Int. J. Eng. Technol. (IJET) **6**, 544–555 (2014)
8. Pan, Z., Eigenmann, R.: Fast and effective orchestration of compiler optimizations for automatic performance tuning. In: Proceedings of the International Symposium on Code Generation and Optimization, pp. 319–332. IEEE Computer Society (2006)
9. Kulkarni, P.A., Whalley, D.B., Tyson, G.S., Davidson, J.W.: Exhaustive optimization phase order space exploration. In: Proceedings of the International Symposium on Code Generation and Optimization (2006)
10. Cooper, K.D., Schielke, P.J., Subramanian, D.: Optimizing for reduced code space using genetic algorithms. In: ACM SIGPLAN Notices, pp. 1–9 (1999)
11. Triantafyllis, S., Vachharajani, M., Vachharajani, N., August, D.I.: Compiler optimization-space exploration. In: Proceedings of the International Symposium on Code Generation and Optimization. Feedback-Directed and Runtime Optimization, 204–215, March 2003
12. Kulkarni, P.A., Hines, S.R., Whalley, D.B., Hiser, J.D., Davidson, J.W., Jones, D.L.: Fast and efficient searches for effective optimization-phase sequences. ACM Trans. Archit. Code Optim. **2**, 165–198 (2005)
13. Guiffaut, C., Mahdjoubi, K.: A parallel FDTD algorithm using the MPI library. IEEE Antennas Propag. Mag. **43**, 94–103 (2001)
14. Idrees, S.K., Fanfakh, A.B.M.: Performance and energy consumption prediction of randomly selected nodes in heterogeneous cluster. In: Al-mamory, S.O., Alwan, J.K., Hussein, A.D. (eds.) NTICT 2018. CCIS, vol. 938, pp. 21–34. Springer, Cham (2018). https://doi.org/10.1007/978-3-030-01653-1_2
15. Ashouri, A.H., Bignoli, A., Palermo, G., Silvano, C., Kulkarni, S., Cavazos, J.: MiCOMP: mitigating the compiler phase-ordering problem using optimization sub-sequences and machine learning. ACM Trans. Archit. Code Optim. (TACO) **14**(3), 29 (2017)

# Using Machine Learning to Predict
# the Sequences of Optimization Passes

Laith H. Alhasnawy, Esraa H. Alwan, and Ahmed B. M. Fanfakh[✉]

Department of Computer Science, College of Science for Women, University of Babylon,
Hillah, Iraq
lythhamd071@gmail.com,
{wsci.israa.hadi,wsci.ahmed.badri}@uobabylon.edu.iq

**Abstract.** The manual tuning for the sequence of optimization passes in modern
compilers was impractical, where this sequence was not general to all benchmark
programs in achieving optimal performance. Therefore, the process of selecting a
set of passes manually them over a particular program to achieve optimal performance is a very difficult problem. Moreover, choosing the order for these passes
will add another problem called phase order problem.

In this paper, the proposed approach provides auto tuning optimization
sequences instead of manual tuning by building a prediction scheme. The proposed framework used machine learning to find a sequence for each program by
collecting these passes based on the features of program. K-Nearest Neighbors
classifier algorithm (KNN) is used in the prediction process and improved by that
the reduction algorithm that work after it. The reduction algorithm eliminated
passes that have a negative impact on program execution time.

The proposed approach was evaluated using the LLVM (Low Level Virtual
Machine) compiler under Linux Ubuntu. The obtained results showed that this approach outperforms standard optimization level-O2 of LLVM compiler in improving the execution time byan average of 21% through building a prediction scheme
by using KNN algorithm. Consequently, the execution time is improved byan
average of 23% due to the application of the reduction algorithm on the sequences
of optimization passes resulting from KNN algorithm.

**Keywords:** LLVM · KNN algorithm · Reduction algorithm · Optimization
sequence · Optimization passes

## 1 Introduction

Modern compilers are portioned into three parts: the front end, middle end and back end.
The frontend creates an intermediate representation (IR) for the source program after
testing its syntactic and semantic rightness. The middle end which represents the optimizer is a critical part. It applies a sequence of conversions on the IR in order to optimize
it to get better execution time, code size, and power consumption. The backend converts
the optimized IR to target assembly code. The optimizer part arranged as a sequence of
optimization passes that represents analysis passes followed by transformation passes.

© Springer Nature Switzerland AG 2020
A. M. Al-Bakry et al. (Eds.): NTICT 2020, CCIS 1183, pp. 139–156, 2020.
https://doi.org/10.1007/978-3-030-55340-1_10

Analysis passes analyses the program and fetches important information needed through the transformation passes [1].

Compiler developers work on the optimization passes to produce an optimized version from the original version. A code segment may be a function, a basic block, a source file, and the whole program. However, code optimizations are based on application, programming language, and architecture. Generally, the number of optimizations passes in the LLVM compiler is more than 100 passes, while the GCC compiler has more than 200 passes where they are organized at standard optimization levels (e.g. -O1, -O2, -O3) [2].

Every set of optimization passes that can work on the program IR is named an optimization sequence. The set of all optimization sequences is called the optimization sequence space where it is infinitely large because the probability to producing optimization sequences has been increased with growing the number of optimization passes. Thus, the number of the optimization sequences equals $(k)^L$, where k represents the number of optimizations passes, and L represents the optimization sequence length. However, there is a problem in finding a good optimization sequences because there are a large number of optimization sequences that contain many passes that interact with each other in complex ways [1] and [5].

These optimization sequences are applied in a fixed order. This order is tuned manually for a specific group of programs impractically. It also requires to tuned it again at every time the compiler is modified to a new version or when a new optimization passes has been added [8]. Therefore, the word "optimization" is an inaccurate name because these compilers may have a good performance for particular programs, but its performance is bad for other programs, see [2] and [12]. Moreover, there is a problem in understanding the effect of the optimization passes behavior on the code segment to be optimized. These due to these passes are interacted with each other in an unpredictable manner during the optimization process. Therefore, there are many problems in the choosing the passes that achieve the optimal performance for a particular program and in the choosing the order for these passes [2] and [11].

To solve these problems, this paper introduces a method for auto tuning optimization sequences by building a prediction scheme using machine learning. This method predicts the sequences of optimization passes for each program and order these passes based on the features of that program through applying K-Nearest Neighbors classifier algorithm (KNN).

KNN is one of the machine learning algorithms that work under supervised learning. This algorithm deals with classification tasks in the wide applications of various fields such as pattern recognition, cluster analysis for image databases, internet marketing, and etc. It is proposed by Fix & Hodges, and it is called KNN algorithm because it relies on the use of the nearest neighbors to predict the class of unknown data belonging to those neighbors. It is classified the unknown data based on its features to the nearest neighbor, nearest neighbor is calculated based on the k value that determines the number of nearest neighbors that have been observed after applying the similarity measures [3] and [10].

Then, another algorithm called a reduction algorithm is applied to eliminate optimization passes that have a negative impact on program execution time from the resulting optimization sequence by using KNN algorithm.

This paper is organized as follow: Sect. (2) describes the related works to the proposed approach. Sect. (3) illustrates the proposed approach and the steps of building a prediction scheme. Sect. (4) shows the experimental results of the proposed approach. Finally, Sect. (5) describes the conclusion.

## 2 Literature Review

In this section, some of the related works that has been done in this field are listed. Matrins et al. [4] presented a Design Space Exploration scheme that uses a clustering approach to find groups of good optimization sequences. This approach reduces search space for the number of optimization passes used during the exploration of optimization sequences, where it combines the previously suggested optimization passes into each group to reduce the exploration time and improve the performance. The unseen program is compiled with different groups of optimization sequences, and the best optimization sequence is chosen for it.

Cavazos et al. [6] proposed a machine learning approach depend on a logistic regression scheme to find good optimization sequence for a particular program. This approach used performance counters which based on the dynamic features for each program when executed, where a similarity metric is found between any two programs depending on the similarities between these features. A machine learning approach predicted mechanism depending on the behavior of many of the trained programs in a training phase to create several optimization sequences. The obtained sequences used for training programs, and the prediction mechanism in the test phase to predict the optimization sequence that will be allowed to compile the unseen program by a training phase.

Purini and Jain [1] proposed an approach to find good optimization sequences sets, which are able to cover several programs in each set. This approach uses random and genetic algorithms to create multiple effective sets of optimization sequences and eliminate passes that cause a negative impact from each set.

Ashouri et al. [7] presented a machine learning approach using Bayesian Network to handle the problem of finding the best optimization sequence. This Network provides a probability distribution for the optimization sequences based on the dynamic features extracted for programs. These features feed the network to produce a program-specific which bias to the best optimization sequence from the probability distribution.

Kulkarni and Cavazos [8] applied a Neuro-Evolution depends on a machine learning approach to build an artificial neural network to predict best optimization sequence for program to be improved. This network uses input features that describe the current state of the program, where the best optimization sequence is found by applying network on all passes to determine the probability of these passes. Then the features are extracted again after applying this network one more time on the pass which represent the highest probability because it represents the best pass prediction in the network. Thus, this process continues for several iterations until the best optimization sequence is found from the successful passes for previous attempts which have been applied to the network.

Junior et al. [15] proposed a hybrid approach to solve the optimization sequence selection problem, which combines machine learning using Support Vector Machine and iterative compilation using Genetic Algorithm. This approach is initially used machine learning to identify potential optimization sequences depending on programs characteristics. Then, it uses iterative compilation to adapt these potential optimization sequences to explore potential search optimization sequences and returns the best optimization sequence by reducing the search space for these potential optimization sequences, this procedure is done through a so-called solution adapter which uses a strategy depends on the iterative compilation to improve the solution found in machine learning. Authors in [16], used multi selection genetic algorithm that work over a cluster of programs to find the best optimization sequence for each cluster.

## 3 Proposed Approach

The main objective of the proposed approach is to provide auto tuning optimization sequences by building a prediction scheme. This scheme uses machine learning to find a sequence for each program instead of the manual tuning. The outline of the proposed approach is as follows:

- Extract static features for each program before and after applying passes, where each pass is applied to all programs.
- Calculate the execution time for each program before and after applying passes.
- Using KNN algorithm to build a prediction scheme.
- Applying reduction algorithm to improve the resulting sequences from KNN.

The LLVM Clanguage frontend (Clang) converts the source program code to machine readable bit code file format (.bc). Moreover, it is providing two useful tools, the first one is opt, and the second is llc. The first tool optimizes the program by applying the specific sequence of optimization passes and stores the result in bit code format file. The second tool converts an LLVM IR from a bit code file to targetcode. After compiling the program using Clang -O0 level (i.e. no optimization applied) then the optimization pass called–scalarrepl (scalar replacement) is applied alongside with each pass. This pass converts the LLVM IR to Static Single Assignment (SSA) to enable other passes to perform their optimizations. Now, the specific optimization sequence for each program can be applied depending on its features by using the opt tool as well. Finally, llc tool converts the LLVM IR produced from the optimizer to target code. Table 1 shows the optimization passes used in -O2 level of LLVM compiler [1].

**Table 1.** Optimization passes of –O2 standard optimization level

| List of optimizations passes | | | |
|---|---|---|---|
| -scalarrepl | -always-inline | -argpromotion | -codegenprepare |
| -constmerge | -constprop | -correlated-propagation | -dce |
| -deadargelim | -die | -dse | -early-cse |
| -globaldce | -globalopt | -gvn | -indvars |
| -inline | -instcombine | -instsimplify | -internalize |
| -ipconstprop | -ipsccp | -jump-threading | -licm |
| -loop-deletion | -loop-idiom | -loop-instsimplify | -loop-reduce |
| -loop-rotate | -loop-simplify | -loop-unroll | -loop-unswitch |
| -loops | -lower-expect | -loweratomic | -lowerinvoke |
| -lowerswitch | -memcpyopt | -mergefunc | -mergereturn |
| -partial-inliner | -prune-eh | -reassociate | -adce |
| -sccp | -simplify-libcalls | -simplifycfg | -sink |
| -tailcallelim | -targetlibinfo | -no-aa | -tbaa |
| -basicaa | -basiccg | -functionattrs | -scalarrepl-ssa |
| -domtree | -lazy-value-info | -lcssa | -scalar-evolution |
| -memdep | -strip-dead-prototypes | | |

Figure 1 shows a summary of the compilation process for each program in our approach by using the optimization passes shown in the Table 1 when building a prediction scheme.

## 3.1 Features Extraction Using Instruction Count

To extract the static features, the programs must be stored in bit code file format (.bc). Then LLVM InstCount pass is applied to extract them. Overall, these features describe the static behavior for program. Table 2 illustrates the LLVM static features. The total number of them equals 39 features distributed among different programs. Each one has its own numerical value, which it is varying from one program to another as shown in [13] and [14].

In our approach these features are used as shown below:

1. In the training set, the features for each program are extracted before and after applying each pass to each program.
2. When using KNN algorithm, the similarity for each program is calculated based on its features as shown in the scheme of finding optimization sequence.
3. In the testing set, each program is rounded based on its features to the closest program in the training set by using KNN algorithm.

There are many similarity measures to calculate the similarity for features such as Cosine, Simple Matching Coefficient, Jaccard Coefficient and etc. The cosine similarity shown in Eq. (1) was used as a similarity measure in our approach because it is one of

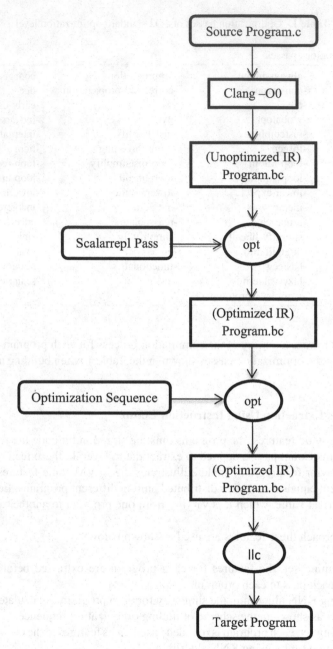

**Fig. 1.** Program compilation process.

**Table 2.** The static features for programs

| List of static features | | | | | |
|---|---|---|---|---|---|
| Add instructions | FAdd instructions | GetElementPtr instructions | Ret instructions | SRem instructions | ZExt instructions |
| Alloca instructions | FCmp instructions | ICmp instructions | SDiv instructions | Shl instructions | basic blocks |
| And instructions | FDiv instructions | Load instructions | Sub instructions | Store instructions | Memory instructions |
| AShr instructions | FMul instructions | Mul instructions | Switch instructions | Trunc instructions | non-external functions |
| BitCast instructions | FPExt instructions | or instructions | SExt instructions | URem instructions | |
| Br instructions | FPToSI instructions | PHI instructions | Select instructions | Unreachable instructions | |
| Call instructions | FSub instructions | PtrToInt instructions | SIToFP instructions | Xor instructions | |

the most useful measures when using sparse data which contain a few non-zero values (i.e. asymmetric). This measure ignores the zero values matching and handles other non-zero values [9].

$$\text{Sim}(p, pi) = \frac{\sum_{w=1}^{m} Pw \times Piw}{\sqrt{\sum_{w=1}^{m} (Pw)^2} \times \sqrt{\sum_{w=1}^{m} (Piw)^2}} \qquad (1)$$

Where $p$ represents the basic program, and $pi$ represents other programs. In our approach we observe this sparse because "for example" there is a program that has 15 features from the total number that equals 39 features, and another program has 25 features from the total number. This difference in the features number between these two programs, and also the difference for these two programs with the total number of features are processed by placing zero value for features not available in these programs in order to be equal to the total number of features. Thus, each one of the programs contains 39 values mostly zero. As a result, most feature values are zero for the benchmark programs which have been processed through this metric.

### 3.2 Build a Prediction Scheme

In our approach, the prediction scheme is built to find the sequence of optimization passes for each program through the data set as shown below.

### 3.2.1   Data Set

The data set consists of programs taken from several benchmarks, see [12] for more information [12]. Examples about these benchmarks are polybench, shootout, standford, and etc. Benchmark programs include various topics like image processing, data structures, data mining, tail recursive, linear algebra. The data set is divided into 70% of the training set and 30% of the testing set.

#### 3.2.1.1 Training Set

In the training set, the features for each training program are extracted before and after applying each pass, and the execution time is also calculated before and after applying each pass. Whcrc - scalarrepl pass was applied together with each pass.

After, KNN algorithm is applied to find a sequence of optimization passes for each program. Then the order of these passes is determined according to the program features.

As a result, these programs are represented as a tree consisting of 62 programs divided into 5 levels, where each level contains twice the level that precedes it (i.e. the first level contains 2 programs, the second level on 4 programs, the third level on 8 programs, the fourth level on 16 programs, and the fifth level on 32 programs) as shown in Fig. 2.

To find the sequence of optimization passes for each program in training set, the similarity is calculated between its feature (before applying any pass) and other program features. Then, two of the most similar programs ($K = 2$) are chosen to the original one and add their passes to the optimization sequence (SequencArray as shown in Algorithm 1). Each one of chosen program is considered as a new program, then it classified again to the nearest two neighbor programs in the tree by using its features (after applying the pass). Another two similar programs are chosen where their execution time after applying optimization passes represents the lowest one. This process will be repeated until all the programs in the training set are chosen.

#### 3.2.1.2 Testing Set

In the testing set, another 20 benchmark programs are fetched as unseen programs. Then, classify each program depending on its features to the closest program in the training set by using KNN algorithm to predict the good optimization sequence for this program, which represents the same optimization sequence for the closest program in the training set.

### 3.3   Interaction Between Optimization Passes

After finding the optimization sequences for all the benchmark programs in prediction scheme, we noticed that there are some the optimization passes in the generated sequences have a negative impact on program execution time due to the wrong interactions between these passes.

To eliminate the optimization passes that cause the wrong interactions, the proposed approach uses an algorithm called the reduction algorithm. This algorithm inspects the effect of optimization passes on program execution time and check their impact to determine whether it good or bad [1]. Therefore, a good optimization pass that improves program execution time remains in the resulting optimization sequence, but the bad pass that does not improve the program execution time is removed from that sequence.

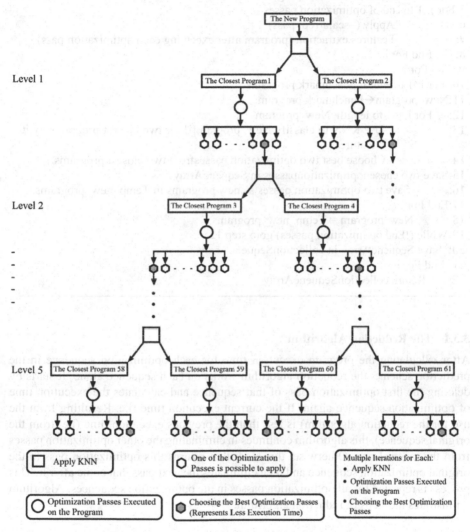

**Fig. 2.** The scheme of finding optimization sequence

---

**Algorithm 1: Finding Optimization Sequence**

**Input:** benchmark programs, optimization passes.
**Output:** optimization sequence of passes for each program (*collectionSequenceArray*).
1: For i=1 to end of benchmark programs
2:      Features extraction (program before executing optimization passes)
3: End For i
4: For i=1 to end of benchmark programs
5:For j=1 to end of optimization passes
6:           Apply (–scalarrepl pass)
7:           Features extraction (program after executing each optimization pass)
8:      End For j
9: End For i
10: For i=1 to end of benchmark programs
11:New_program← benchmark program
12:    For j = 1 to length New_program
13:           Use KNN to classify New_program[j] for two closest programs by its
               features.
14:           Choose best two optimization passesfrom two closest programs.
15:Save two chosenoptimizationpasses in SequencArray.
16:       Save two optimization passes as new programs in Temp_new_programs.
17:End For j
18           New_program = Temp_new_programs
19:While (!End optimization passes) goto step 12
20:  Save SequencArray in collectionSequenceArray
21: End For i
22:     Return collectionSequenceArray

---

### 3.3.1  The Reduction Algorithm

After calculating the program execution time for each optimization sequence in the prediction scheme, the reduction algorithm works on each sequence alone. It starts by deleting the first optimization pass of that sequence and calculates the execution time of optimization sequence again. If the current execution time (i.e. Resulting from the use for the reduction algorithm) is less than the previous execution time (i.e. from the original sequence), this algorithm continues in eliminating the other optimization passes from the original one. Otherwise, this algorithm returns this optimization pass to the original optimization sequence and then eliminates the next pass that come after it. This process is repeated for all optimization passes in the optimization sequences. Algorithm 2 illustrates the steps of reduction algorithm.

---

**Algorithm 2:Reduction Optimization Sequence**

Input:prog_Opt_seq by using KNN.

Output: Best_Opt_seq.

1: Execution time ← execution (prog_Opt_seq)

2: I← 1

3: While I< length of prog_Opt_seq

5:        Temp_prog_Opt_seq← Delete (prog_Opt_seq, I)

6:        Best_Execution time ← execution(Temp_prog_Opt_seq)

7:    If Best_Execution time < Execution time

8:        Execution time ← Best_Execution time

9:Best_Opt_seq←Temp_prog_Opt_seq

10:I←0

11:End if

12:        I←I+1

13: End  while

14:        Return  Best Opt_seq

---

## 4   Experimental Results

Our approach was evaluated on Intel (R) Core (TM) i5 processor with 3.7 GB RAM running a Linux - Ubuntu 16.4 operating system.

Each program from the benchmark programs was executed with its own optimization sequence by using the script file. Each program has been executed three times and the average is calculated for them to increase the accuracy of the results.

In this section, the results will be displayed for 10 benchmark programs as a sample of the training set and 10 benchmark programs as a sample of the testing set. Tables 3 and 4 show the optimization sequences obtained for the programs of these samples in our approach.

Figures 3 and 4 show the execution time obtained for the programs shown in Tables 3 and 4.

**Table 3.** The optimization sequence before and after the reduction process in the training set.

| Programs | Optimization sequence before the reduction process | Number of passes | Optimization sequence after the reduction process | Number of passes |
|---|---|---|---|---|
| Training program1.c | -instcombine -inline -loop-rotate -early-cse -indvars -gvn -basicaa -basiccg -prune-eh -loop-idiom -loop-reduce -jump-threading -tailcallelim -loop-deletion -licm | 15 | -instcombine -inline -early-cse -indvars -gvn -basicaa -basiccg -prune-eh -loop-idiom -loop-reduce -jump-threading -tailcallelim -loop-deletion -licm | 14 |

*(continued)*

**Table 3.**  (*continued*)

| Programs | Optimization sequence before the reduction process | Number of passes | Optimization sequence after the reduction process | Number of passes |
|---|---|---|---|---|
| Training program2.c | -early-cse -instcombine -basicaa -basiccg -prune-eh -loop-rotate -loop-idiom -gvn -loop-reduce -inline -jump-threading -tailcallelim -indvars -loop-deletion -licm | 15 | -early-cse -instcombine -basicaa -basiccg -prune-eh -loop-rotate -loop-idiom -gvn -loop-reduce -jump-threading -tailcallelim -indvars -loop-deletion -licm | 14 |
| Training program3.c | -prune-eh -loop-rotate -tailcallelim -indvars -early-cse -loop-deletion -loop-idiom -gvn -jump-threading -inline -basicaa -loop-reduce -basiccg -instcombine -licm | 15 | -prune-eh -tailcallelim -indvars -early-cse -loop-deletion -loop-idiom -gvn -jump-threading -inline -basicaa -loop-reduce -basiccg -instcombine -licm | 14 |
| Training program4.c | -basiccg -gvn -early-cse -loop-rotate -prune-eh -inline -basicaa -instcombine -loop-idiom -jump-threading -loop-reduce -tailcallelim -indvars -licm -loop-deletion | 15 | -basiccg -gvn -early-cse -loop-rotate -prune-eh -basicaa -instcombine -loop-idiom -jump-threading -loop-reduce -tailcallelim -indvars -licm -loop-deletion | 14 |
| Training program5.c | -loop-idiom -loop-rotate -basicaa -gvn -early-cse -basiccg -prune-eh -inline -jump-threading -loop-reduce -instcombine -tailcallelim -indvars -licm -loop-deletion | 15 | -loop-idiom -loop-rotate -basicaa -gvn -early-cse -basiccg -prune-eh -inline -jump-threading -loop-reduce -instcombine -indvars -licm -loop-deletion | 14 |
| Training program6.c | -tailcallelim -inline -prune-eh -indvars -gvn -jump-threading -early-cse -loop-rotate -loop-deletion -loop-idiom -basicaa -loop-reduce -basiccg -instcombine -licm | 15 | -tailcallelim -inline -prune-eh -indvars -gvn -jump-threading -early-cse -loop-rotate -loop-deletion -loop-idiom -basicaa -loop-reduce -basiccg -instcombine -licm | 15 |

(*continued*)

**Table 3.** (*continued*)

| Programs | Optimization sequence before the reduction process | Number of passes | Optimization sequence after the reduction process | Number of passes |
|---|---|---|---|---|
| Training program7.c | -loop-rotate -gvn -prune-eh -early-cse -jump-threading -loop-idiom -basicaa -inline -basiccg -tailcallelim -indvars -loop-deletion -loop-reduce -instcombine -licm | 15 | -loop-rotate -gvn -prune-eh -basicaa -inline -basiccg -tailcallelim -indvars -loop-deletion -loop-reduce -instcombine -licm | 12 |
| Training program8.c | -inline -tailcallelim -loop-rotate -basicaa -gvn -loop-reduce -instcombine -early-cse -loop-idiom -prune-eh -basiccg -jump-threading -indvars -loop-deletion -licm | 15 | -inline -tailcallelim -loop-rotate -basicaa -gvn -loop-reduce -early-cse -loop-idiom -basiccg -indvars -loop-deletion -licm | 12 |
| Training program9.c | -loop-idiom -loop-rotate -early-cse -gvn -inline -basiccg -prune-eh -jump-threading -loop-reduce -tailcallelim -basicaa -instcombine -indvars -loop-deletion -llcm | 15 | -loop-idiom -loop-rotate -early-cse -gvn -inline -basiccg -prune-eh -jump-threading -loop-reduce -tailcallelim -basicaa -instcombine -indvars loop-deletion -licm | 15 |
| Training program10.c | -loop-reduce -gvn -early-cse -loop-idiom -basicaa -loop-rotate -inline -prune-eh -basiccg -jump-threading -tailcallelim -instcombine -indvars -loop-deletion -licm | 15 | -loop-reduce -gvn -early-cse -loop-idiom -basicaa -loop-rotate -inline -prune-eh -basiccg -jump-threading -tailcallelim -instcombine -indvars -loop-deletion –licm | 15 |

According to the Tables 3 and 4, there are some programs that do not apply the reduction process on their optimization sequences because there are no wrong interactions in passes of those optimization sequences when applying the reduction algorithm. Therefore, these optimization sequences remain the same after the reduction process for those programs.

The results mentioned in Figs. 3 and 4 showed the comparison at the execution time when using both the KNN algorithm and the KNN algorithm with the reduction algorithm in the prediction to the standard optimization level -O2.

These results indicate that our approach outperforms standard optimization level-O2 by improving the execution time. This though building a prediction scheme to find a good sequence of optimization passes for each program from the benchmark programs

**Table 4.** The optimization sequence before and after the reduction process in the testing set.

| Programs | Optimization sequence before the reduction process | Number of passes | Optimization sequence after the reduction process | Number of passes |
|---|---|---|---|---|
| Testing program1.c | -prune-eh -early-cse loop-rotate -loop-idiom -basicaa -basiccg -gvn -inline -jump-threading -loop-reduce -tailcallelim -instcombine -indvars -loop-deletion -licm | 15 | -prune-eh -early-cse -loop-rotate -loop-idiom -basicaa -basiccg -gvn -inline -jump-threading -loop-reduce -tailcallelim -instcombine -indvars -loop-deletion -licm | 15 |
| Testing program2.c | -early-cse -inline -prune-eh -loop-idiom -loop-rotate -basicaa -gvn -jump-threading -basiccg -loop-reduce -instcombine -tailcallelim -indvars -loop-deletion -licm | 15 | -early-cse -inline -prune-eh -loop-idiom -loop-rotate -basicaa -gvn -basiccg -instcombine -indvars -loop-deletion | 11 |
| Testing program3.c | -loop-rotate -basicaa -early-cse -gvn -loop-idiom -inline -prune-eh -basiccg -jump-threading -loop-reduce -tailcallelim -instcombine -indvars -loop-deletion -licm | 15 | -loop-rotate -basicaa -early-cse -gvn -loop-idiom -inline -prune-eh -basiccg -jump-threading -loop-reduce -tailcallelim -instcombine -indvars -loop-deletion -licm | 15 |
| Testing program4.c | -loop-reduce -indvars -early-cse -loop-idiom -loop-rotate -inline -prune-eh -basicaa -basiccg -gvn -instcombine -tailcallelim -jump-threading -licm -loop-deletion | 15 | -loop-reduce -indvars -early-cse -loop-idiom -loop-rotate -inline -prune-eh -basicaa -basiccg -gvn -instcombine -tailcallelim -jump-threading -licm -loop-deletion | 15 |

(*continued*)

**Table 4.** (*continued*)

| Programs | Optimization sequence before the reduction process | Number of passes | Optimization sequence after the reduction process | Number of passes |
|---|---|---|---|---|
| Testing program5.c | -loop-idiom -loop-rotate -early-cse -gvn -inline -basiccg -prune-eh -jump-threading -loop-reduce -tailcallelim -basicaa -instcombine -indvars -loop-deletion -licm | 15 | -loop-idiom -loop-rotate -early-cse -gvn -inline -basiccg -prune-eh -jump-threading -loop-reduce -tailcallelim -basicaa -instcombine -indvars -loop-deletion -licm | 15 |
| Testing program6.c | -prune-eh -loop-rotate -basicaa -inline -gvn -early-cse -loop-idiom -basiccg -jump-threading -loop-reduce -tailcallelim -instcombine -indvars -loop-deletion -licm | 15 | -prune-eh -loop-rotate -basicaa -inline -gvn -early-cse -loop-idiom -basiccg -jump-threading -loop-reduce -tailcallelim -instcombine -indvars -loop-deletion -licm | 14 |
| Testing program7.c | -loop-reduce -loop-rotate -early-cse -loop-idiom -basicaa -inline -prune-eh -basiccg -jump-threading -gvn -tailcallelim -instcombine -indvars -loop-deletion -licm | 15 | -loop-reduce -loop-rotate -early-cse -loop-idiom -basicaa -prune-eh -basiccg -gvn -tailcallelim -instcombine -indvars -loop-deletion -licm | 13 |
| Testing program8.c | -loop-rotate -basicaa -early-cse -gvn -loop-idiom -inline -prune-eh -basiccg -jump-threading -loop-reduce -tailcallelim -instcombine -indvars -loop-deletion -licm | 15 | -loop-rotate -basicaa -early-cse -gvn -loop-idiom -inline -prune-eh -basiccg -jump-threading -loop-reduce -tailcallelim -instcombine -indvars -loop-deletion -licm | 15 |

(*continued*)

**Table 4.** (*continued*)

| Programs | Optimization sequence before the reduction process | Number of passes | Optimization sequence after the reduction process | Number of passes |
|---|---|---|---|---|
| Testing program9.c | -loop-reduce<br>-loop-rotate<br>-early-cse<br>-loop-idiom -basicaa<br>-inline -prune-eh<br>-basiccg<br>-jump-threading<br>-gvn –tailcallelim<br>-instcombine<br>-indvars<br>-loop-deletion -licm | 15 | -loop-reduce<br>-loop-rotate<br>-early-cse<br>-loop-idiom -basicaa<br>-prune-eh -basiccg<br>-gvn -tailcallelim<br>-instcombine<br>-indvars<br>-loop-deletion -licm | 13 |
| Testing program10.c | -gvn -early-cse<br>-loop-rotate -basicaa<br>-loop-idiom -inline<br>-prune-eh -basiccg<br>-jump-threading<br>-tailcallelim<br>-instcombine<br>-loop-reduce<br>-indvars<br>-loop-deletion -licm | 15 | -early-cse<br>-loop-rotate -basicaa<br>-loop-idiom -inline<br>-prune-eh -basiccg<br>-jump-threading<br>-tailcallelim<br>-instcombine<br>-loop-reduce<br>-indvars<br>-loop-deletion -licm | 14 |

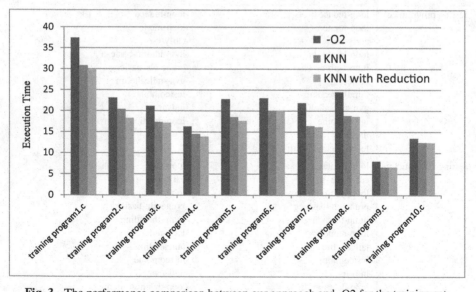

**Fig. 3.** The performance comparison between our approach and -O2 for the training set.

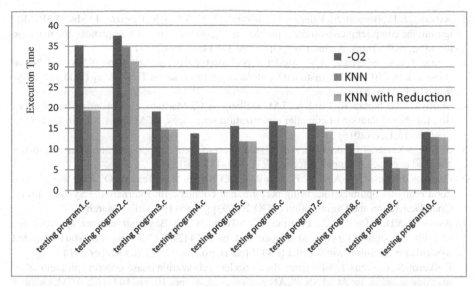

**Fig. 4.** The performance comparison between our approach and -O2 for the testing set.

based on its features using KNN algorithm. The results of application only KNN show an improvement ratio of 21% on average in the execution time. There is also an increase in improving the execution time when applying the reduction algorithm on the sequences of optimization passes by 23% on average of the execution time compared with KNN without reduction.

## 5 Conclusion

This work presented machine learning to build a framework prediction scheme for compiler optimization sequence. This scheme provided auto tuning to solve the problems resulting from manual tuning. The KNN algorithm was used in the prediction scheme to find the sequence of optimization passes for each program from the benchmark programs, and then order these passes based on the features of that program. The reduction algorithm was also applied to eliminate passes that have a negative impact on program execution time for the resulting sequence by using KNN algorithm.

   LLVM infrastructure has been used to validate the proposed method. The experiment results showed that the proposed approach outperforms LLVM -O2 by improving the execution time by average of 21% when building a prediction scheme using KNN algorithm. Moreover, it improved the execution time byan average of 23% after applying reduction algorithm a long side to the KNN algorithm.

## References

1. Purini, S., Jain, L.: Finding good optimization sequences covering program space. ACM Trans. Archit. Code Opt. (TACO) **9**(4), 56 (2013)

2. Ashouri, A.H., Bignoli, A., Palermo, G., Silvano, C., Kulkarni, S., Cavazos, J.: MiCOMP: Mitigating the compiler phase-ordering problem using optimization sub-sequences and machine learning. ACM Trans. Archit. Code Opt. (TACO) **14**(3), 29 (2017)
3. Amra, I.A.A., Maghari, A.Y.: Students performance prediction using KNN and Naïve Bayesian. In: 2017 8th International Conference on Information Technology (ICIT), pp. 909–913. IEEE, May 2017
4. Martins, L.G., Nobre, R., Cardoso, J.M., Delbem, A.C., Marques, E.: Clustering-based selection for the exploration of compiler optimization sequences. ACM Trans. Archit. Code Opt. (TACO) **13**(1), 8 (2016)
5. de Souza Xavier, T.C., da Silva, A.F.: Exploration of compiler optimization sequences using a hybrid approach. Comput. Inform. **37**(1), 165–185 (2018)
6. Cavazos, J., Fursin, G., Agakov, F., Bonilla, E., O'Boyle, M.F., Temam, O.: Rapidly selecting good compiler optimizations using performance counters. In: International Symposium on Code Generation and Optimization (CGO 2007), pp. 185–197. IEEE, March 2007
7. Ashouri, A.H., Mariani, G., Palermo, G., Silvano, C.: A Bayesian network approach for compiler auto-tuning for embedded processors. In: 2014 IEEE 12th Symposium on Embedded Systems for Real-time Multimedia (ESTIMedia), pp. 90–97. IEEE, October 2014
8. Kulkarni, S., Cavazos, J.: Mitigating the compiler optimization phase-ordering problem using machine learning. In: ACM SIGPLAN Notices, vol. 47, no. 10, pp. 147–162. ACM, October 2012
9. Tan, P.N., Steinbach, M., Kumar, V., et al.: Introduction to Data Mining (2006)
10. Suguna, N., Thanushkodi, K.: An improved k-nearest neighbor classification using genetic algorithm. Int. J. Comput. Sci. Issues **7**(2), 18–21 (2010)
11. Pan, Z., Eigenmann, R.: Fast and effective orchestration of compiler optimizations for automatic performance tuning. In: Proceedings of the International Symposium on Code Generation and Optimization, pp. 319–332. IEEE Computer Society, March 2006
12. Alkaaby, Z.S., Alwan, E.H., Fanfakh, A.B.M.: Finding a good global sequence using multi-level genetic algorithm. J. Eng. Appl. Sci. **13**, 9777–9783 (2018)
13. kiran_c: Counting Instructions with LLVM. https://kiran-c.livejournal.com/7956.html
14. Criswell, J.: Need some feed back. https://marc.info/?l=cfe-dev&m=130935922030121 &w=2
15. Junior, N.L.Q., Rodriguez, L.G.A., da Silva, A.F.: Combining machine learning with a genetic algorithm to find good complier optimizations sequences. In: ICEIS, vol. 1, pp. pp. 397–404 (2017)
16. Almohammed, M.H., Alwan, E.H., Fanfakh, A.B.M.: Programs features clustering to find optimization sequence using genetic algorithm. In: Jain, Lakhmi C., Peng, S.-L., Alhadidi, B., Pal, S. (eds.) ICICCT 2019. LAIS, vol. 9, pp. 40–50. Springer, Cham (2020). https://doi.org/10.1007/978-3-030-38501-9_4

# A Proposed Method for Feature Extraction to Enhance Classification Algorithms Performance

Aysar J. Hassooni[1(✉)], Mohammed Abdullah Naser[1], and Safaa O. Al-Mamory[2]

[1] College of Science for Women, University of Babylon, Babylon, Iraq
acerjaber@gmail.com, wsci.mohammed.abud@uobabylon.edu.iq
[2] College of Business Informatics, University of Information Technology and Communications, Baghdad, Iraq
salmamory@uoitc.edu.iq

**Abstract.** Classification is one of the main tasks of machine learning. A proposed method to improve the accuracy of classification by adding new six features to the original dataset is proposed in the paper. These new six features are extracted from supervised and unsupervised set of algorithms. The decisions of three different classification algorithms and three different clustering algorithms are added as new features. These algorithms are selected with a condition of having various techniques to build models and in order to guarantee diversity. Moreover, three scenarii are proposed to generate the new features. Then, simple voting algorithm is applied to the new dataset (i.e. original dataset and new added features). The experimental results on 10 different datasets from UCI repository and NSL-KDD dataset proved that the proposed features ranked better than original features and the accuracy enhanced by about 12%.

**Keywords:** Classification · Clustering · Multi-classifier system · Ensemble · Feature extraction

## 1  Introduction

It is obvious that significant enhancements can be achieved in challenging machine learning problems through combining the opinions of multiple classifiers. In spite of combining several algorithms produce improved performance than using the same algorithm with a different set of parameters [1]. However, a better ensemble detection rate could be obtained by balancing accuracy and diversity [2]. Three principal factors should be considered while building an ensemble classifier which are: the selected algorithms, the organization of these algorithms, and the combination rules [3]. The organization of a multiclassifier could be classified as either modular (i.e. the training set is unique for each classifier) or ensemble (each classifier works on the same state space).

The proposed system accomplishes diversity not by traditional comparing the algorithms' output with the class attribute, but by using a supervised and unsupervised set of algorithms together. In spite of using an ensemble set of classifiers is not a new task;

© Springer Nature Switzerland AG 2020
A. M. Al-Bakry et al. (Eds.): NTICT 2020, CCIS 1183, pp. 157–166, 2020.
https://doi.org/10.1007/978-3-030-55340-1_11

however, the important contribution of this paper is to mix the decisions of different machine learning algorithms (clustering and classification) with the set of features in prediction. In other words, unlike other ensemble algorithms, the original features will contribute in the final decision.

In this paper, a Hybrid Combination Rule (HCR) for multiple classifier system is proposed by combining a different set of classifiers and clustering algorithms with the original set of features. In other words, a set of classifiers are trained independently on the same dataset, then the decisions of these classifiers are added as new features together with the original features. In addition, another set of clustering algorithms are applied on the same dataset and the clustering results are added as new features. Finally, a new classifier is taking the decision depending on the old and new set of features. The experimental results on 11 different datasets proved that the proposed model outperforms classical voting algorithm.

This paper is organized as follows. Section 2 will cover related works. Section 3 covers the major part of the new proposed system and Sect. 4 includes the empirical side and the results obtained, respectively. The conclusions were covered in Sect. 5.

## 2  Related Works

There are many research contributions related to ensemble classification. Some of the tremendous applications, researches and ideas of ensemble classification in recent years are listed below:

A. Dzelihodzic et al. studied the effectiveness of several algorithms to classify credit applications in the case of a single classifier as compared to their use as ensemble techniques. The single classification algorithms are neural networks, decision trees, transport support machine machines, while the AdaBoost and Bagging are ensemble technologies. For model validation, K-fold cross-validation is used. An experiment is performed on the dataset of Bosnian commercial bank and the results indicate that both Bagging and AdaBoost ensemble methods have better classification accuracy than single classifiers. The use of ensemble methods also contributes to better evaluation of apps for loans [4].

Fadi Salo et al. suggested an ensemble feature selection method along with an anomaly detection technique combining clustering and classification machine learning methods to classify the network traffic to recognize patterns of attack previously unseen. Three distinct techniques of feature selection, namely gain information, correlation, and magnitude that are used as part of an ensemble model selecting eight prevalent features. Firstly, k-Means clustering based Manhattan distance is used to partition the training instances into k clusters. Support vector machines, nearest k neighbors, random forests, and quadratic discriminating analysis are used as a classification model based on the resulting clusters, representing a density region of normal or anomaly cases. The classifier efficiency assessed using the Kyoto dataset. The experimental results indicate that the suggested framework is effective in identifying earlier invisible patterns of assault compared to the traditional approach to classification [5].

P. Trajdos et al. used their score functions to present a combined linear classifier. The scoring function's value relies on the distance from the decision boundary. Two score functions were tested and investigated four distinct combination approaches. The primary objective of this job is to determine the best combination approach for the geometric space in question. The suggested method has been implemented to the heterogeneous ensemble and has been contrasted with two techniques of reference, which are the majority vote and model average. The comparison was created with seven distinct quality criteria. The outcome achieved shows that the best combination methods of the geometric mixture are combination strategies based on a straightforward average and trimmed average [6].

Gang Wang et al. conducted a comparative evaluation of the performance of three learning methods of the ensemble; these methods are bagging, boosting and random subspace based on five learners, namely Naive Bayes, maximum entropy, decision tree, k-nearest neighbor and vector support. Using 10-fold cross-validation for their job, they randomly chose ten sentiment analysis datasets. Based on a total of 1200 comparison group studies, their experimental results indicated that ensemble techniques enhance efficiency for sentiment classification by individual base learners [7].

Yongjun Piao et al. proposed an ensemble method to classify high-dimensional information, with each classifier being built from a different set of features determined by partitioning redundant features. The redundancy of features in their method is considered to divide the original space of the feature. Then, a support vector machine trains each generated feature subset and the results of each classifier are combined by majority vote. The proposed method showed that it was efficient compared with other ensemble techniques and that the obtained results were superior to other methods [8].

Amine Bayoudhi et al. attempted to enhance the results of document classification in Arabic sentiment analysis by combining distinct types of features such as characteristics of view and discourse; and by suggesting an ensemble-based classifier to explore its contribution to Arabic sentiment classification. The results of this study showed a macro-averaged F-measure achievement of 85.06% and showed that discourse features enhanced F-measure slightly by about 3% or 4% [9].

Nazmin Sultana et al. assessed the efficiency of ensemble techniques on two distinct datasets in combination with the meta-algorithm. Also, the majority voting algorithm used to predict the classification of sentiment for the proposed model. Seven common supervised machine learning algorithms (bagging, boosting & stacking) were selected for text classification and ensemble approach. The selected machine learning algorithms are naïve Bayes (multinomial and Bernoulli), logistical regression, supporting vector machine, stochastic gradient descent, neighbor k-nearest, and perceptron multi-layer. The results demonstrate that in some instances the strategy of meta-classifier-based ensemble outperforms individual classifier [10].

E. Fersini et al. produced a Bayesian ensemble method that takes advantage of various classifiers to predict user-generated content sentiment orientation and enhance the efficiency of polarity classification activities. They discussed the issue of classifier choice by suggesting a greedy strategy that assesses each model's contribution to the ensemble. The experimental results on the gold standard data set showed that the proposed method was more efficient than the traditional classification and ensemble method [11].

## 3  The Proposed System

In the proposed system, a new method is suggested to improve the performance of the classifiers by employing classification (supervised) and clustering (unsupervised) algorithms in features' generation. This is accomplished through the process of creating or extracting new features added to the original data and then applying the new data to other classification algorithms in order to increase the accuracy of the classifying of this data. The new extracted features make use of the diversity techniques of supervised and unsupervised algorithms in features extraction. The block diagram of HCR is given in Fig. 1.

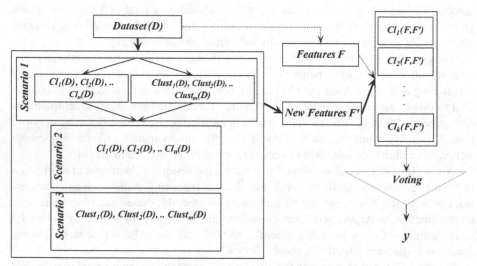

**Fig. 1.** Proposed system for a hybrid combination rule (HCR) for multiple classifier system

The proposed system uses three scenarii to generate the new features. The first scenario assume that the supervised (i.e. $Cl_1(D)$, $Cl_2(D)$, ... $Cl_n(D)$) and unsupervised (i.e. $Clust_1(D)$, $Clust_2(D)$, ... $Clust_m(D)$) algorithms work on the original dataset. It is assumed that the number of new generated features is $n+m$. Moreover, the set supervised algorithms used in different scenarii and the set of supervised algorithms used in voting are disjoint. The second scenario assumes the supervised algorithms work on the original dataset to generate $n$ new features which is added to the original dataset. The third scenario is different from the second scenario in using the unsupervised algorithms. The three proposed scenarii is stated in Fig. 1. The proposed workflow appears in the figure in bold arrows while the standard voting algorithm appears in dotted arrows.

Every machine learning algorithm has a unique intelligent technique to represent the state space. This intelligent representation could be included in the dataset as new features. These new features add more diversity to the learning process. Moreover, the voting ensemble has been used with whole set of features to increase classification performance. The voting equation for the proposed system is appeared in Eq. 1 where

$N_c$ is the highest number of votes for instance t.

$$y = argmax(N_c(Cl_1^t(F, F')), N_c(Cl_2^t(F, F')), \ldots, N_c(Cl_n^t(F, F')))  \quad (1)$$

The main characteristics of the HCR system include the following. Firstly, it is easy to use because it depends on a set of commonly used machine learning algorithms. Secondly, the use of several classifiers is better in covering a larger number of problems. Thus, this system will be more reliable than using a single classifier for classification. Finally, it utilizes the principles of supervised and unsupervised algorithms together making the classification prediction better.

## 4  Experimental Results

A series of experiments are conducted to test the efficiency of the system with several datasets from UCI repository [13] and NSL-KDD dataset [14] having different number of classes, and different number of features as shown in Table 1. Holdout method to split these datasets into train and test is used in all experiments. Moreover, several programs had been used like weka 3-9-3 [15] and java programming language to get results from these datasets. All results obtained using a Dell system core™i3-2370 M, RAM 4G windows 32-bit.

**Table 1.**  The used dataset in the experiments from UCI repository and NSL-KDD

| No. | Dataset | Instances | Features | Classes |
|-----|---------|-----------|----------|---------|
| 1 | Diabetes | 768 | 8 | 2 |
| 2 | Ionosphere | 351 | 34 | 2 |
| 3 | Parkinson | 197 | 16 | 2 |
| 4 | Phishing | 11,055 | 31 | 2 |
| 5 | Iris | 150 | 4 | 3 |
| 6 | Lung cancer | 32 | 56 | 3 |
| 7 | User Knowledge Modeling (UKM) | 258 | 5 | 4 |
| 8 | Glass | 214 | 9 | 6 |
| 9 | Ecoli | 336 | 7 | 8 |
| 10 | Yeast | 1,484 | 8 | 10 |
| 11 | NSL-KDD | 148,517 | 48 | 40 |

Four measures are used in all experiments to test proposed system performance. These measures include accuracy, precision recall, and F-measure where these measures depend on computing confusion matrix [16]. These measures are appeared in Eq. 2 to Eq. 5. Comparison results of HCR with scenario 1 are presented in Table 2. HCR with scenario 1 outperforms voting method for all dataset except for Ionosphere, Parkin sons, and User-modeling datasets. However, the results (on average) of HCR with scenario 1 are better than voting. It should be noted that the voting system uses three base classifiers in all experiments in this paper which are SVM, J48, and Naïve Bayes from Weka libraries. The reason for using these three base classifiers is the great diversity found in the principles of the work of these three algorithms. Moreover, HCR with scenario 1 uses three supervised algorithms (which are KNN, Logistic Regression (LOG), and Random Forest (RF)) and three unsupervised algorithms (i.e. K-Means clustering, Hierarchical clustering (HC), and Self-organized Map clustering (SOM)).

$$Accuracy = (TP + TN)/(TP + TN + FP + FN) \tag{2}$$

$$Precision = TP/(TP + FP) \tag{3}$$

$$Recall = TPR = TP/(TP + FN) \tag{4}$$

$$F - measure = 2 * \left( \frac{Precision * Recall}{Precision + Recall} \right) \tag{5}$$

**Table 2.** The comparison of the results between the voting method and the HCR system with scenario 1

| Dataset | Voting | | | | HCR with scenario 1 | | | |
|---|---|---|---|---|---|---|---|---|
| | Accuracy % | Precision | Recall | F-Measure | Accuracy% | Precision | Recall | F-Measure |
| Diabetes | 77.82 | 0.77 | 0.77 | 0.77 | 87.15 | 0.88 | 0.87 | 0.86 |
| E.coli | 66.96 | – | 0.67 | – | 78.57 | 0.82 | 0.78 | 0.78 |
| Glass | 66.19 | 0.68 | 0.66 | 0.65 | 80.28 | – | 0.8 | – |
| Ionosphere | 96.58 | 0.96 | 0.96 | 0.96 | 95.72 | 0.95 | 0.95 | 0.95 |
| Iris | 88 | 0.93 | 0.88 | 0.89 | 98 | 0.98 | 0.98 | 0.98 |
| Lung-cancer | 45.45 | 0.45 | 0.45 | 0.44 | 54.54 | 0.55 | 0.54 | 0.54 |
| Parkinsons | 83.07 | 0.84 | 0.83 | 0.81 | 80 | 0.82 | 0.8 | 0.78 |
| Phishing | 94.7 | 0.94 | 0.94 | 0.94 | 97.31 | 0.97 | 0.97 | 0.97 |
| UKM | 88.37 | 0.92 | 0.88 | 0.88 | 70.93 | – | 0.7 | – |
| Yeast | 54.14 | 0.57 | 0.54 | 0.54 | 78.58 | 0.82 | 0.78 | 0.79 |
| NSL-KDD | 68.69 | – | 0.68 | – | 72.15 | – | 0.72 | – |
| Average | 75.45 | 0.78 | 0.75 | 0.76 | 81.2 | 0.84 | 0.8 | 0.83 |

Table 3 shows the obtained results after application of the HCR system with scenarii 2 and 3. Some results cannot be obtained from Weka therefore dash is replaced these values in the table. It can be noted from the tables that the accuracy of HCR with scenario 3 outperforms voting (on average), scenario 1, and scenario 2 by about 12%, 6%, and 8% respectively. The accuracy improvement ratio is low when using supervised algorithms with the HCR system compared with the use of unsupervised algorithm or uses both them. Moreover, the precision of HCR with scenario 3 outperforms voting (on average), scenario 1, and scenario 2 by about 0.12, 0.6, and 0.9 respectively. The other measures are also enhanced.

**Table 3.** The results obtained from HCR with scenario 2 and scenario 3.

| Dataset | HCR with scenario 2 | | | | HCR with scenario 3 | | | |
|---|---|---|---|---|---|---|---|---|
| | Accuracy% | Precision | Recall | F-Measure | Accuracy % | Precision | Recall | F-Measure |
| Diabetes | 80.54 | 0.8 | 0.8 | 0.8 | 100 | 1 | 1 | 1 |
| E.coli | 71.42 | 0.76 | 0.71 | 0.68 | 93.75 | 0.95 | 0.93 | 0.93 |
| Glass | 74.64 | 0.75 | 0.74 | 0.74 | 92.95 | – | 0.93 | – |
| Ionosphere | 94.01 | 0.94 | 0.94 | 0.94 | 98.29 | 0.98 | 0.98 | 0.98 |
| Iris | 96 | 0.96 | 0.96 | 0.96 | 100 | 1 | 1 | 1 |
| Lung-cancer | 54.54 | 0.55 | 0.54 | 0.54 | 45.45 | 0.455 | 0.455 | 0.44 |
| Parkinsons | 80 | 0.82 | 0.8 | 0.78 | 100 | 1 | 1 | 1 |
| Phishing | 97.31 | 0.97 | 0.97 | 0.97 | 94.73 | 0.94 | 0.94 | 0.94 |
| UKM | 93.02 | 0.94 | 0.93 | 0.93 | 67.44 | – | 0.67 | – |
| Yeast | 58.18 | 0.59 | 0.58 | 0.57 | 96.36 | 0.95 | 0.96 | 0.95 |
| NSL-KDD | 72.29 | – | 0.72 | – | 71.95 | – | 0.72 | – |
| Average | 79.26 | 0.81 | 0.79 | 0.79 | 87.35 | 0.9 | 0.87 | 0.9 |

On the other hand, the HCR system reduces the rate of Root Mean Square Error (RMSE) [17]; RMSE is the standard deviation of the residuals (prediction errors). Residuals are a measure of how far from the regression line data points are; RMSE is a measure of how to spread out these residuals are. In other words, it tells you how concentrated the data is around the line of best fit. Figure 2 shows RMSE rate for voting method and the HRC system. It could be noted that the RMSE value in most cases is better than voting. The RMSE of HCR with scenario 3 is less than voting (on average), scenario 1, and scenario 2 by 0.1479, 0.0399, and 0.0312, respectively. In other words, the features extracted from unsupervised algorithms reduced the error rate.

As another measure to find the meaningful of the new six extracted features, feature ranking algorithms are used. Weka software is used to rank the features by using Gain Ratio Attribute Evaluator and Ranker search method. The rank of each feature for every dataset is presented in Table 4. The number of features appeared in the table is the original number increased by six (i.e. the number of new features). The extracted features ranked as first feature for 11 datasets and second for also 11 datasets and third for seven datasets. In addition, the extracted features are ranked from the best 7% to 40% on average.

A comparison with the state of the art ensemble algorithms is presented in Table 5. These algorithms include Bagging, Adaboost, and voting. The same measures are used in the comparison which includes accuracy, precision recall, and F-measure. It can be noted from the table that the accuracy of HCR with scenario 3 outperforms voting

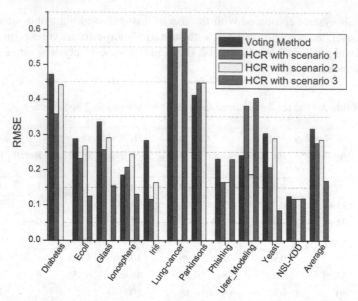

**Fig. 2.** RMSE for the voting system and the HCR system with different scenarii

**Table 4.** The ranking for the new features

| Dataset | #features | Supervised features | | | Unsupervised features | | | Rank average |
|---------|-----------|---------------------|---------------------|------------------|-----------------------|------------------|------------------|--------------|
| | | Feature 1 KNN | Feature 2 LOG | Feature 3 RF | Feature 4 K-means | Feature 5 HC | Feature 6 SOM | |
| Diabetes | 14 | 4th | 5th | 3rd | $1^{st}$ | 2nd | 9th | 4 |
| E.coli | 13 | 4th | 5th | 2nd | $7^{th}$ | 1st | 9th | 5 |
| Glass | 15 | 8th | 7th | 2nd | $4^{th}$ | 1st | 6th | 5 |
| Ionosphere | 40 | 4th | 2nd | 1st | 29 | 3rd | 39 | 13 |
| Iris | 10 | 4th | 3rd | 2nd | $7^{th}$ | 1st | 6th | 4 |
| Lung-cancer | 62 | 8th | 1st | 2nd | 44 | 53 | 14 | 20 |
| Parkinson | 29 | 3rd | 5th | 2nd | 24 | 1st | 25 | 10 |
| Phishing | 37 | 2nd | 3rd | 1st | 16 | – | 24 | 9 |
| UKM | 11 | 4th | 3rd | 2nd | $1^{st}$ | 6th | 7th | 4 |
| Yeast | 14 | 5th | 6th | 3rd | $2^{nd}$ | 1st | 9th | 4 |
| NSL-KDD | 54 | 2nd | – | 1st | 23 | – | 19 | 4 |

(on average), Bagging, and Adaboost by about 12%, 20.62%, and 19% respectively. Moreover, all other measures are enhanced.

**Table 5.** A comparison made with state of the art ensemble systems

| System | Accuracy | Precision | Recall | F-Measure |
|---|---|---|---|---|
| Bagging | 66.73 | – | 0.66 | – |
| Adaboost | 68.25 | – | 0.67 | – |
| Voting | 75.45 | 0.78 | 0.75 | 0.76 |
| HCR with scenario 1 | 81.2 | 0.84 | 0.8 | 0.83 |
| HCR with scenario 2 | 79.26 | 0.81 | 0.79 | 0.79 |
| HCR with scenario 3 | 87.35 | 0.9 | 0.87 | 0.9 |

## 5 Conclusions

A new method was used to improve the accuracy of the classification of the data when using several classifiers by implementing several algorithms (classification and clustering) on the features and extracted new features and then adding these new features to the original features and the implementation of a set of other classification algorithms (base classifiers) on the original and new features. We have noted the accuracy of the classification increases by (12%) when adding the new features resulting from the clustering algorithms, and also increases it by(4.23%) when added when adding the new features resulting from the classification algorithms and finally the accuracy increases by (10.75%) when adding the features resulting from both. We conclude from this proposed method that clustering algorithms have greater influence in improving the accuracy from classification algorithms.

## References

1. Schubert, E., Wojdanowski, R., Zimek, A., Kriegel, H.-P.: On evaluation of outlier rankings and outlier scores. In: Proceedings of the 12th SIAM International Conference on Data Mining (SDM), Anaheim, CA, 2012, pp. 1047–1058 (2012)
2. Zimek, A., Campello, R.J., Sander, R.: Ensembles for unsupervised outlier detection: challenges and research questions a position paper. SIGKDD Explor. Newsl. **15**, 11–22 (2014)
3. Canuto, A.M., Abreu, M.C., de Melo Oliveira, L., Xavier, J.C., Santos, A.D.M.: Investigating the influence of the choice of the ensemble members in accuracy and diversity of selection-based and fusion-based methods for ensembles. Pattern Recogn. Lett. **28**, 472–486 (2007)
4. Dželihodžić, A., Đonko, D.: Comparison of ensemble classification techniques and single classifiers performance for customer credit assessment. Model Artif. Intell. **11**(3), 140–150 (2016)
5. Salo, F., Injadat, M., Moubayed, A., Nassif, A.B., Essex, A.: Clustering enabled classification using ensemble feature selection for intrusion detection. In: 2019 International Conference on Computing, Networking and Communications (ICNC), pp. 276–281. IEEE (2019)
6. Trajdos, P., Burduk, R.: Combination of linear classifiers using score function – analysis of possible combination strategies. In: Burduk, R., Kurzynski, M., Wozniak, M. (eds.) CORES 2019. AISC, vol. 977, pp. 348–359. Springer, Cham (2020). https://doi.org/10.1007/978-3-030-19738-4_35

7. Wang, G., et al.: Sentiment classification: the contribution of ensemble learning. Decis. Support Syst. **57**, 77–93 (2014)
8. Piao, Y., et al.: A new ensemble method with feature space partitioning for high-dimensional data classification. Math. Probl. Eng. (2015)
9. Bayoudhi, A., Ghorbel, H., Belguith, L.H.: Sentiment classification of Arabic documents: experiments with multi-type features and ensemble algorithms. In: Proceedings of the 29th Pacific Asia Conference on Language, Information and Computation, pp. 196–205 (2015)
10. Sultana, N., Islam, M.M.: Meta classifier-based ensemble learning for sentiment classification. In: Uddin, M.S., Bansal, J.C. (eds.) Proceedings of International Joint Conference on Computational Intelligence. AIS, pp. 73–84. Springer, Singapore (2020). https://doi.org/10.1007/978-981-13-7564-4_7
11. Fersini, E., Messina, E., Pozzi, F.A.: Sentiment analysis: Bayesian ensemble learning. Decis. Support Syst. **68**, 26–38 (2014)
12. Phyu, T.N.: Survey of classification techniques in data mining. In: Proceedings of the International MultiConference of Engineers and Computer Scientists, vol. 1, pp. 18–20 (2009)
13. UCI Machine Learning Repository. http://www.ics.uci.edu/~mlearn/MLRepository.html
14. NSL-KDD dataset. https://www.unb.ca/cic/datasets/nsl.html
15. Hall, M., Frank, E., Holmes, G., Pfahringer, B., Reutemann, P., Witten, I.H.: The WEKA data mining software: an update. SIGKDD Explor. Newsl. **11**(1), 10–18 (2009)
16. Davis, J., Goadrich, M.: The relationship between Precision-Recall and ROC curves. In: Proceedings of the 23rd 123 International Conference on Machine Learning, pp. 233–240. ACM (2006)
17. Wang, W., Lu, Y.: Analysis of the mean absolute error (MAE) and the root mean square error (RMSE) in assessing rounding model. In: IOP Conference Series: Materials Science and Engineering, vol. 324, no. 1, p. 012049. IOP Publishing (2018)

# Heart Disease Prediction System Using Optimization Techniques

Namariq Ayad Saeed[✉] and Ziyad Tariq Mustafa Al-Ta'i[✉]

College of Science, Diyala University, Diyala, Iraq
gemy91gemy@gmail.com, ziyad1964tariq@gmail.com

**Abstract.** One of the major causes of death throughout the world is heart disease as estimated by the World Health Organization (WHO). It cannot be easily detected by the medical practitioners as it a difficult task that demands higher knowledge and expertise for prediction. Some computational approaches were suggested for heart disease predication. This study proposes a binary particle swarm optimization algorithm combined with mutual information filter for feature selection and employed logistic regression for classification. The Cleveland heart disease dataset emloyed. Out of a total of 297 instances of patient data, 60 of them employed for testing, and 237 used for training. The results showed 98.33% accuracy of logistic regression obtained with attributes selected by mutual information combined with binary particle swarm optimization algorithm but with attributes chosen from the binary particle swarm optimization algorithm achieved 96.66%.

**Keywords:** Heart disease · Feature selection · Particle swarm optimization algorithm · Mutual information · Optimization algorithm

## 1 Introduction

The heart is an important part of the human body. Which pumps blood to the whole body, If the circulation of the blood in the body is inefficient the members like a brain suffers and if the heart stops working altogether, death happens within a few minutes. Life is entirely based on the efficient functioning of the heart. The term Heart disease indicates the disease of the heart and blood vessel system within it [1].

Cardiovascular disease recorded by the World Health Organization (WHO) as one of the highest contributing factors to human deaths in all the years. Who estimated that in the year 2012, 56 million people lost their lives and the highest contributing factor was heart disease which took lives of 17.5 million peoples, which is representing 31% of all deaths in worldwide. Of these deaths, an estimated 6.7 million caused by stroke, and 7.4 million caused by coronary heart disease. Due to heart disease, almost 23.6 million people will die in 2030 as estimated by the World Health Organization [2].

Healthcare collects a large amount of data which can be in various formats like charts, numbers, text, and images but unfortunately, this information rarely used for clinical decision making. Feature selection is known as a primary preprocessing step in data mining for detecting relevant subset for classification. The high dimensionality of

© Springer Nature Switzerland AG 2020
A. M. Al-Bakry et al. (Eds.): NTICT 2020, CCIS 1183, pp. 167–177, 2020.
https://doi.org/10.1007/978-3-030-55340-1_12

the data can cause a different problem such as reducing the accuracy and increasing the complexity, i.e., "curse of dimensionality". The aim of attribute selection is affording faster construction of the prediction model with better efficiency [3].

Clinical decisions are often decided based on the doctor's intuition and experience instead of the knowledge-rich data hidden in the database. This practice causes excessive medical costs, unwanted basics that affect the quality of service given to patients [4].

The suggested method removes irrelevant attributes to accurately predicts heart disease.

## 2 Related Work

In previous works, different authors have employed various techniques on heart dataset collected from Cleveland Heart Disease database, the performance of the classifier evaluated, and their outputs are analyzed. Some of the authors have implemented classification methods like Naïve Bayes, Artificial Neural Network, K-nearest neighbor, Decision tree, etc., on the heart dataset and different feature selection algorithm such as cuckoo search algorithm and genetic algorithm, etc., then comparisons of the efficiency of various techniques will be implemented.

In [5]: proposed a system for coronary heart diseases predication using the development of a Neuro-genetic model. They implemented attribute subset selection by employing a multi-objective genetic algorithm without any loosing in the accuracy of ANN, which used as heart disease predictor. The efficiency of the developed model evaluated by a database from the Cleveland Clinic Foundation. This study attained high testing accuracy of 89.58% through minimized feature subset.

In [6]: proposed an enhanced system for prediction of heart disease by using a genetic algorithm as a feature selection to select the optimal subset. Radial Basis Function (RBF) Network which is a type of neural network is used to create the classifier applied on UCI Cleveland dataset, the results show that in comparison with Naïve Bayes and J48 RBF Network attained the highest accuracy of 83.83% with all features and 85.48% with reduced subset.

In [7], submitted a novel attribute selection technique for prediction of heart disease with less number of attributes. The dataset used is collected from UCI Cleveland heart disease database. Several data mining algorithm applied for selected subset such as random forest, logistic regression, SMO, neural network, and decision tree. An accuracy of 93.1% achieved with a neural network.

In [8]: proposed a genetic algorithm (GA) with trained recurrent fuzzy networks for heart disease diagnosis using UCI Cleveland data set as having heart disease or not. The biases and weights of the RFNN were decoded as 64 bits long genes for the GA, and the output of GA used to train RFNN. The result showed that 97.78% accuracy.

In [9]: submitted Dragonfly Algorithm (DA) for the weights of every connection between the ANN neurons in a model known as (ANN-DA) for a medical prediction on Cleveland dataset attained an accuracy of 90.21%.

But none of researchers above experiment two stage for feature selection by using feature ranking as first stage then applying one of the optimization algorithm on the output features before applying the classification algorithm.

# 3 Dataset and Metholodgy

## 3.1 Dataset

The Cleveland dataset (UCI, 1990) which employed for this study obtained from the University of California Irvine (UCI) for heart disease dataset that consists of four independent datasets. It includes 303 instances of patient data, but 6 of them have missing values. Cleveland dataset attributes illustrated in Table 1 with their definitions [10].

**Table 1.** UCI Cleveland dataset

| NAME | Definition |
|------|-----------|
| Age | Age expressed in years |
| Cp | Indicates type of chest pain |
| Sex | Sex |
| Trestbps | blood pressure in mm Hg |
| Fbs | Indicates fasting blood sugar status if is larger than 120 mg/dl or not |
| Chol | Serum cholestoral in mg/dl |
| Thalach | Maximum heart rate attained |
| Restecg | Resting electrocardiographic results |
| Exang | Exercise caused by angina |
| Slope | The peak exercise slope |
| Odpeak | ST depression caused by exercise relative to rest |
| Thal | Status of heart |
| Ca | Indicates number of major vessel (0-3) which colored by flourosopy |
| Num | Heart diseases diagonsis (0 = healthy; 1 = Sick1; 2 = Sick2; 3 = Sick3; 4 = Sick4) |

In this study, Cleveland heart disease dataset containing six instances of missing entries omitted. The diagnosis of heart disease feature (num) categorized into two classes referred to the presence (num = 1 or 2 or 3 or 4) and absence (num = 0) of the heart disease. Class distributions indicated as 46% having heart disease and 54% absence. The dataset partioned into two subsets for training (237 instances) employed and for testing (60 instances).

## 3.2 Mutual Information (MI)

MI is a filter method that measures the dependence between the stochastic variables. The primary idea of employing MI in attribute selection is the attributes must highly correlate with the target class. Where MI between two random variables A = (a1, a2 ... a3) and B = (b1, b2 ... b3) [11]:

$$MI(A, B) = \sum_e \sum_f p(a, b) \log \frac{p(a, b)}{p(a)p(b)} \tag{1}$$

Where p(b) and p(a) are indicating the marginal probability distribution for A and B. If A and B are closely related to each other, the MI between them will be high and vise versa.

### 3.3 Naïve Bayes

The NB algorithm is a probabilistic classifier which computes a set of probabilities by counting the combinations of values and frequency in a dataset. It employs Bayes theorem and supposes all features to be independent given the class value. That conditional independence assumption rarely happens in real-world applications, hence the description as Naïve yet the algorithm tends to do great and learn rapidly in different supervised classification issues [12].

NB assume that all attributes make a role in classification and are mutually correlated. That hypothesis is called class conditional independence [13]. The Naïve Bayes Classifier depending on Bayes Theorem. Bayes theorem illustrated in Eq. (2):

$$P(H|X) = \frac{P(X|H)P(H)}{P(X)} \tag{2}$$

Where H is some hypothesis, such that data of X belongs to particular class C, X is some evidence defined by measure on a set of variables, $(H|X)$ is indicating the posterior probability which the hypothesis H holds based on the evidence X, $P(H|X)$ is the posterior probability of X which is conditioned on H., P(H) is indicating the prior probability of H and is independent on X.

Muhammad, Triyanna, and Prasetya, Gauss density function employed in classification of continuous data as shown in Eq. (3)

$$P(X_i = x_i | Y = y_j) = \frac{1}{\sqrt{2\pi}\sigma_{ij}} e^{\frac{(x_i - \mu_{ij})^2}{2\sigma_{ij}^2}} \tag{3}$$

While P is an probability, xi is the value of predictor variable, Xi is a predictor variable, yi is sub-class of Y, $\pi$ indicates a constant value of 3.14.

### 3.4 Logestic Regression

Logistic regression is considered a kind of regression analysis in statistics employed for prediction of the outcome of independent variables or a categorical dependent variable (which can have a bounded number of values) from a set of predictors. The dependent variable always consists of two categories (binary) in logistic regression. Logistic regression is primarily employed for prediction and computing the success probability. It is employing the Eq. 4 to the data [14]:

$$Y = \beta_0 + \beta_1 x_1 + \beta_2 x_2 + \ldots + \beta_n x_n \tag{4}$$

Where the ß$_{0,1,..n}$ are the model parameters, $x_{1,...,n}$ are training samples.

The maximum likelihood estimation usually used in regression coefficients evaluation. Its ratio can help to specify the independent variables statistical significance on the dependent variables and estimates the contribution of individual predictors. The probability (p) of every case is computed by employing the odds ratio,

$$\frac{P}{1-P} = e^{Y} - Y \tag{5}$$

Where p is the probability of every case, y is evaluated from Eq. 4.

### 3.5 Particle Swarm Optimization (PSO)

Kennedy and Eberhart in 1995 proposed PSO, which is an evolutionary computation technique. PSO simulates social behavior like bird's flocking and fish schooling. The fundamental concept of PSO is that knowledge enhanced from the social interaction in in the population not only by the individual experience [15].

A population, also known as a swarm of a candidate solutions, which encoded in the search space as particles. PSO begins with a stochastic initialization of a swarm of particles. The entire swarm moves in the search space to find the best solution by modifying the position of every particle depending upon its own and its neighboring particles experience [16].

Each particle i have a position shown by $x_i = (x_{i1}, x_{i2}, ..., x_{iD})$ and a velocity is illustrated by $v_i = (v_{i1}, v_{i2}, ..., v_{iD})$ that is bounded by a predefined maximum velocity, $v_{max}$ and $v_{id}^t \in [-v_{max}, v_{max}]$, where D is the dimensionality of the search space. Every particle owns a memory to save the position where it found its best experience and expressed as pbest. The global best understood as the best encounter among all particles, expressed as gbest. For every particle the velocity and position modified. Both pbest and gbest can help in modifying the velocity and position depending on the next equations [17].

$$x_{id}(t+1) = x_{id}(t) + v_{id}(t+1) \tag{6}$$

$$v_{id}(t+1) = w * v_{id}(t) + c_1 * r_1 * (p_{id} - x_{id}(t)) + c_2 * r_2 * (p_{gd} - x_{id}(t)) \tag{7}$$

### 3.6 Binary Particle Swarm Optimization (BPSO)

Kennedy and Eberhart proposed a binary form of the particle swarm algorithm (BPSO) for handling optimization issues in discrete domains. The velocity is still modified using Eq. (7), where $x_{id}$, $p_{id}$ and $p_{gd}$ are bounded to 1 or 0. The velocity in BPSO refers to the probability of the identical element in the position vector having value 1. A sigmoid functions ($v_t$) is introduced to convert $v_t$ to the range of (0, 1). The position of every particle in BPSO version is modified according to the next formulae: [17]

$$x(t+1) = \begin{cases} 1, & \text{if } rand() < S(v(t+1)) \\ 0, & \text{otherwise} \end{cases} \tag{8}$$

where S(v(t)) is the Sigmoid function

$$s(v_t) = \frac{1}{1 + e^{-v_t}} \tag{9}$$

where rand() is a stochastic number picked from a uniform distribution in [0,1].

## 4  Proposed Method

The proposed method employs mutual information to calculate the attribute-class mutual information to reach a maximum relevance between the target class and the selected attributes. This hybrid algorithm consists of a filter and wrapper type for attribute subset selection implemented for enhancing the efficiency of heart disease prediction system. Figure 1 describes the proposed hybrid system.

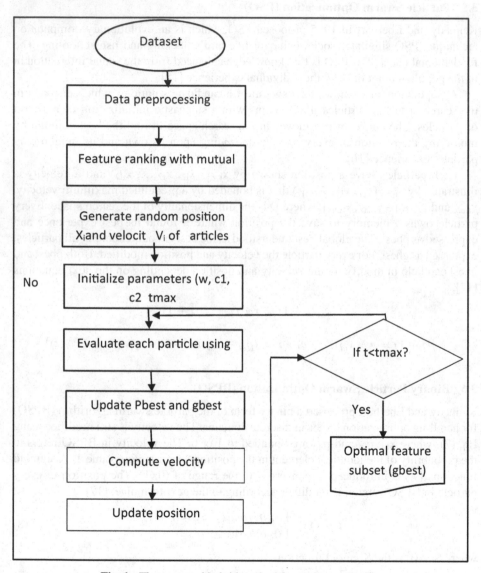

**Fig. 1.** The proposed hybrid method for attribute selection

The attribute selection process implemented in the next steps based on Fig. 1:

Step 1: Implementing mutual information on the dataset and remove the attribute with mutual information equal to zero.

Step 2 (Initialization): The attribute subsets were produced randomly as the initial population of particle swarm optimization (n). Every particle was consist from a string of binary values with a length of m (which represents the number of attributes in the dataset produced from step1), where the value of every bit present the presence of a corresponding attribute in a swarm.

Step 3 (Evaluation): Each particle was estimated by the fitness function. It means that for every particle (attribute subset), the NB model was constructed by training instances in the dataset that contained just attributes in the corresponding attribute subset, and it was assessed by testing instances.

In this step, every particle has a memory to save the position where it find its best experience and referred to pbest. The global best, which is the best find among all particles, denoted as gbest. The gbest and pbest was identified depending on a fitness function.

$$\text{Fitness} = \text{Accuracy of NB algorithm} \tag{10}$$

Step 4 (Updating): Updates the main coefficients. The velocity (V), and position of every particle based on pbest and gbest using Eqs. (8 and 9).

---

Algorithm I. Binary particle swarm optimization combined with Mutual Information (MI_BPSO) Feature selection method

---

**Input:** Dataset.

**Output:** Selected features .

**Steps:**

1. Initialize n (number of particles), $t_{max}$ (maximum number of iteration) , m (number ofattributes),.
2. Initialize t=0, c1, c2,w, $V_{max}$, $V_{min}$, $V_i$ (velocity) for i=1,2, …,n
3. Initialize the fitness of particles fitness(i)=$[0]_{mx1}$
4. Set pbest=0, gbest=$[0]_{1xn}$as the best fitness.
5. **for** each attribute from the main attribute set(d=1,2,…,m)
6.     Compute $MI(f_d ;y)$
7.     If $MI(f_d ;y)>0$
8.         **ffa = ffa** $\cup$ $f_d$

---

9.    end if
**10. end for**
**11.** Intialize $X_i$ (positions), for i=1,2, ...,n.
**12. while** t<$t_{max}$
**13.** for i=1 to n do
**14.** Calculate fitnessof $X_i$ using eq.10
**15.** if fitness[i]>pbest
**16.** xpbest=$X_i$
**17.**        pbest=fitness[i]
**18.**      end if
**19.** if fitness[i]>gbest
**20.** xgbest=$X_i$
**21.**        gbest=fitness[i]
**22.**     end if
**23.**    end for
**24.**    For I =1 To n do
**25.**      For J=1 To m do
**26.**        Calculate $V_{ij}$ using eq.7
**27.**        if $V_{IJ}$> $V_{max}$
**28.** $V_{IJ}$= $V_{max}$
**29.**    end if
**30.**        if $V_{IJ}$< $V_{min}$
**31.**          $V_{IJ}$ = $V_{min}$
**32.**        end if
**33.** Calculate sigmoid function using eq.9
**34.**        Updates $X_i$ using eq.8
**35.** endfor
**36.**    end for
**37.**    t=t+1
**38. End while**
Return xgbest ( as best features subset)

# 5  Experimental Result

## 5.1  Evaluation Criteria

The evaluation of the proposed method was implemented depending on accuracy, recall, precision, and F-measure tests which use the false negative (FN), false positive (FP) true negative (TN), true positive (TP) terms, and these criteria are calculated as follows:

$$precision = \frac{TP}{TP + FP} \tag{11}$$

$$F_{measure} = \frac{2 * TP * precision}{TP + precision} \tag{12}$$

$$recall = \frac{TP}{TP + FN} \tag{13}$$

$$accuracy = \frac{TP + TN}{TP + TN + FP + FN} \qquad (14)$$

## 5.2 Results

In proposed method samples partitioned into two independent data set that randomly selected 80% of the dataset is employed to build and train the classifier while the remaining 20% of the dataset employed for testing.

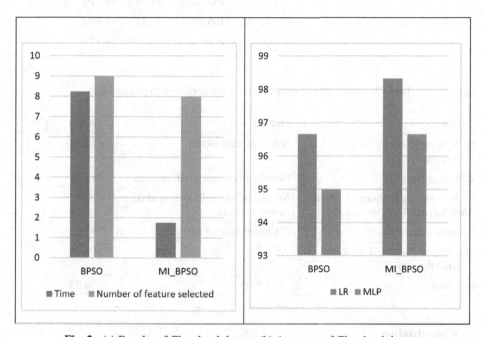

**Fig. 2.** (a) Results of Cleveland dataset (b) Accuracy of Cleveland dataset

First, we employed Z-score normalization to normalize the original data. In the attribute ranking with MI the equation implemented on the dataset and the features equal to zero removed. Based on these Age, Restecg and Fbs were removed. The output of feature ranking model is fed to BPSO. Table 2, Fig. 2 shows the result of proposed system.

Parameters of BPSO are population size is 20, number of iterations is 100, w is set to 0.5, c1 and c2 is set to one, and $V_{max}$ is set to four (Table 3).

**Table 2.** The result of proposed system

| Algorithm | Selected features | Time(s) | Metrics | LR | MLP |
|---|---|---|---|---|---|
| MI_BPSO | 1110110111 | 1.75 | Accuracy | 98.33% | 96.66% |
| | | | Precision | 98.48% | 96.65% |
| | | | Recall | 98.21% | 96.65% |
| | | | F-measure | 98.32% | 96.65% |
| BPSO | 0111011011011 | 8.2343 | Accuracy | 96.66% | 95% |
| | | | Precision | 97.05% | 95.71% |
| | | | Recall | 96.42% | 94.64% |
| | | | F-measure | 96.63% | 94.93% |

**Table 3.** Illustration of comparison of our results with prior studies.

| Author | Methods | Accuracy (%) |
|---|---|---|
| H. S. Niranjana Murthy and M. Meenakshi [5] | Neuro-genetic model | 89.58 |
| A. Durga Devi [6] | Genetic algorithm_RBF network | 85.48 |
| R. Suganya, S. Rajaram, A. Sheik Abdullah and V. Rajendran [7] | Novel attribute selection method and Neural network | 93.1 |
| KaanUyar and AhmetIlhan [8] | GA based trained RFNN | 97.78 |
| MaisYasen, Nailah Al-Madi and Nadim Obeid [9] | ANN-DA | 90.21 |
| Proposed | MI_BPSO | 98.33 |

## 6  Conclusion

In this paper, Mutual Information combined with binary particle swarm optimization algorithm depending on classification system suggested for heart disease prediction. In the proposed approach, MI and BPSO are combined to classify heart disease issues in fast and an efficient manner.

The MI_BPSO consists of two subsystems: Binary particle swarm optimization combined with Mutual Information for feature selection and the classification subsystem using logistic regression and multilayer perceptron. The Cleveland dataset from the UCI machine learning repository was chosen to test the system. The experimental results illustrate that the combination (sex, cp, Restbp, Thalach, exang, Slope, ca, thal) attains the highest classification accuracy (98.33%) with LR and (96.66%) with MLP. The results illustrate that classification accuracy of logistic regression increased from (96.66% to 98.33%) when employing MI_BPSO as attribute selection approach as well as the accuracy of multilayer perceptron increased from (95% to 96.66%). The time algorithm takes decreased from 8.2343 s with binary particle swarm optimization algorithm to 1.75 s with the proposed method. The experiments were implemented on an Intel Core i7, 64-bit Operating System, 4-GB RAM and 2.20-GHz processor.

# References

1. Keerthana, T.K.: Heart disease prediction system using data mining method. Int. J. Eng. Trends Technol. **47**(6), 361–363 (2017)
2. WHO: Fact sheet: Cardiovascular diseases (CVDs). Who (2017)
3. Hsu, H., Hsieh, C., Lu, M.: Expert systems with applications hybrid feature selection by combining filters and wrappers. Expert Syst. Appl. **38**(7), 8144–8150 (2011)
4. Alia, A.F., Taweel, A.: Feature selection based on hybrid binary cuckoo search and rough set theory in classification for nominal datasets. Inf. Technol. Comput. Sci. **4**(April), 63–72 (2017)
5. Murthy, H.S.N., Meenakshi, M.: Dimensionality reduction using neuro-genetic approach for early prediction of coronary heart disease. In: Proceedings International Conference Circuits, Communications Control Computing (I4C 2014), no. November, pp. 21–22 (2014)
6. Devi, A.D., Xavier, S.: Enhanced prediction of heart disease by genetic algorithm and RBF network. Int. J. Adv. Inf. Eng. Technol. **2**(2), 29–37 (2015)
7. Suganya, R., Rajaram, S., Sheik Abdullah, A., Rajendran, V.: A novel feature selection method for predicting heart diseases with data mining techinques. Asian J. Inf. Technol. **15**(8), 1314–1321 (2016)
8. Uyar, K.: Diagnosis of heart disease using genetic algorithm based trained recurrent fuzzy neural networks, vol. 120, pp. 588–593 (2018)
9. Yasen, M., Al-madi, N., Obeid, N.: Optimizing neural networks using dragonfly algorithm for medical prediction (2018)
10. UCI Machine Learning Repository: Heart Disease Data Set. https://archive.ics.uci.edu/ml/datasets/heart+Disease. Accessed 01 Jun 2019
11. Cang, S.: A mutual information based feature selection algorithm, pp. 2241–2245 (2011)
12. Patil, T.R., Sherekar, S.S.: Performance analysis of naive bayes and j48 classification algorithm for data classification. Int. J. Comput. Sci. Appl. **6**(2), 256–261 (2013)
13. Bhavsar, H., Ganatra, A.: A comparative study of training algorithms for supervised machine learning, vol. 2, no. 4, pp. 74–81 (2012)
14. Manninen,T.: Predictive modeling using sparse logistic regression with applications (2014)
15. Xue, B., Zhang, M., Browne, W.N.: Particle swarm optimization for feature selection in classification: a multi-objective approach. IEEE Trans. Cybern. **43**(6), 1656–1671 (2013)
16. Cervante, L., Xue, B., Zhang, M., Shang, L.: Binary particle swarm optimisation for feature selection: a filter based approach. IEEE Congr. Evol. Comput., CEC (2012)
17. Mafarja, M., Jarrar, R., Ahmad, S., Abusnaina, A.A.: Feature selection using binary particle swarm optimization with time varying inertia weight strategies, In: ICFNDS, pp. 1–9 (2018)

# Optimizing Energy in Cooperative Sensing Cognitive Radios

Temitope Fajemilehin[1]($\boxtimes$), Abid Yahya[2], Jaafar A. Aldhaibani[3], and Kibet Langat[4]

[1] Pan African University Institute of Basic Sciences, Technology, and Innovation,
P.O. Box 62000-00200, Nairobi, Kenya
topefajemilehin@gmail.com
[2] Botswana International University of Science and Technology,
Private Bag 16, Palapye, Botswana
yahyabid@gmail.com
[3] University of Information Technology and Communications, Baghdad, Iraq
dr.jaafaraldhaibani@uoitc.edu.iq
[4] Jomo Kenyatta University of Agriculture and Technology, P.O. Box 62000-00200,
Nairobi, Kenya
kibetlp@jkuat.ac.ke

**Abstract.** One of the viable solutions for effective spectrum management is cognitive radio. Single sensing systems are prone to interference; thus, the use of cooperative spectrum sensing. This paper aims to determine the required number of cognitive radios that would optimize the performance of a communication network in terms of energy utilization and bandwidth requirement. The cognitive sensing technique used was energy detection due to its reduced energy, computational, and communication resources requirement. The channel noise variance was set to $-25$ dB. Spectrum sensing was carried out at a frequency of 936 MHz and bandwidth of 200 kHz. Machine learning was first used to enhance the specificity of detection to minimize interference. Genetic Algorithm (GA) and Simulated Annealing (SA) were used to optimize the number of cognitive radios putting into consideration all constraints in the network. Genetic Algorithm gave a better result of two optimization techniques used. It gave an overall reduction of 40.74% in energy conserved without affecting the detection accuracy.

**Keywords:** Cognitive radio · Energy detection · Cooperative spectrum sensing · Optimization · Genetic Algorithm · Simulated Annealing

## 1 Introduction

A considerable section of the radio spectrum (3 GHz–300 GHz region) is not well utilized, and some other parts are heavily crowded due to the growth and need for more telecommunication applications and services [1]. Spectrum holes between 30 MHz and 3 GHz, which are not as occupied, can thus effectively utilized for telecommunication services [2]. This needs to be carried out effectively to avoid fast depletion of the limited

© Springer Nature Switzerland AG 2020
A. M. Al-Bakry et al. (Eds.): NTICT 2020, CCIS 1183, pp. 178–188, 2020.
https://doi.org/10.1007/978-3-030-55340-1_13

electromagnetic resources available with the increase in demand for wireless devices and services [3].

The Radio Communication Sector of the International Telecommunications Union (ITU-R) ensures that the radio spectrum is efficiently allocated to each sector by regulating the spectrum. This helps to prevent interference between different users, especially in this age of wireless technologies demand for increased bandwidth. This leads to an increase in the demand for more telecommunications resources such as bandwidth, power, and communicating frequency slots. Thus, due to the shortage of the useable available radio spectrum, new radio access technologies are therefore limited by. This limitation, which still exists in the present spectrum allocation scheme is due to fixed radio functions, static spectrum allocation, and limited network coordination [4]. Dynamic spectrum management applications is an improvement that is required to accommodate the increased demand for telecommunication resources [5].

Cognitive radio system (CRS) is a dynamic spectrum management technique which is relevant in the white space utilization effort. It is a software-defined radio which can manage the spectrum by exploiting spectrum holes and allowing the deployment of multiple wireless systems [5–7]. It has the capability of sensing the status of the transmitting channels so as to determine the occupancy status of such channels. Also, it can dynamically tune the usage of the spectrum based on factors such as type of radios in the network; bandwidth required allocation, location of the radios, time of the day, etc. [8].

Spectrum sensing is a crucial role in cognitive radio systems (CRS). It is necessary to check if a particular channel is being used by a licensed user at a specific point in time [9]. This is very important to avoid interference and channel misallocation. Energy detection-based sensing is a commonly used spectrum sensing technique. It requires minimal resources to detect the occupancy status of the channel and is not as computationally demanding as the other techniques [10, 11]. However, its limitations include multipath fading, shadowing, and the hidden terminal problem. It also cannot discriminate between the various forms of users utilizing resources on the channel.

The inadequacies of single sensing are compensated for in the cooperative sensing. However, this increases the communication overhead of the network and is usually also not able to determine the type of user occupying the channel, mainly if the energy detection technique is used.

Authors in [12], worked on enhancing the detection sensitivity of energy detection-based spectrum sensing using the concept of an adaptive threshold [12]. Several combining techniques for cognitive radio users in cooperative spectrum sensing were examined for different modulation schemes. Significant improvements were recorded, but the technique was not fully predictable in noisy channels.

Authors in [13], also proposed a method of improvement that enhanced the classical energy detection method and maintained a comparable level of computational complexity and cost. Detection time was also reduced in comparison to other, more sophisticated methods of sensing. This could not distinguish between the different users within the network. In work [14], authors used an augmented spectrum sensing algorithm where the energy detector's detection is augmented by cyclostationary detection. However, the technique needs some knowledge about the primary users' transmission characteristics, which is not always available.

Multiple antenna techniques were used in [12] to improve the performance of energy detection and cyclostationary feature detection-based. This was implemented in a cooperative spectrum sensing scheme using Equal Gain Combining (EGC). The detection sensitivity was improved but was not focused on interference mitigation.

Authors in [15], used different channels without any knowledge of the environment to improve on the usage of the idle spectrum with due consideration for fairness in channel selection. Authors in [16], developed on this with the aid of a p-norm energy detector. The performance of the cooperative spectrum sensing was improved alongside improved gain in κ-μ fading channels. This paper is building upon these improvements with the purpose to increase the capacity of the cognitive radio identify the user occupying the channel.

Authors in [17], used a two-stage reinforcement learning approach to increase cooperative sensing performance. This method minimized the number of sensing operations and reduced the energy required in the sensing operation. The channel sensing and allocation was improved, but required computational resources and learning time.

Reference [18] explored optimal threshold selection at low SNR to step up cognitive radio sensing performance. Better sensing performance was obtained compared to previous approaches, which used an unchanging false-alarm rate and kept the threshold of detection rate constant. However, there was no provision made for instantaneous SNR drop, which sometimes occurs.

Authors in [19], used an adaptive simulated annealing particle swarm optimization (ASASPSO) to improve cognitive radio power allocation. Interference power threshold of the PU, transmission rate limitation of SUs, and the signal to interference and noise ratio (SINR) were the parameters used. Thus, the power consumed was reduced, and an improved SINR and transmission rate was achieved.

In [20], the authors used a dynamic GA to provide an effective spectrum utilization method for PUs and SUs. They aimed at solving the problem of channel allocation in cognitive radio networks. The GA used had a uniform distribution, and the enhanced crossover and mutation functions were used. The algorithm developed by the authors was able to improve the ability of the CR to adapt to a varying environment. The ability to forecast and reduce interference was also enhanced.

Authors in [21], compared the performance of GA and PSO while considering the sensing time, probability of false alarm, cost, iterations, and throughput of the cognitive radio network. The authors considered a trade. The authors had a tradeoff of sensing time over transmission time. GA gave a better ROC curve when compared with PSO when sensing time and throughput was considered. The SNR values were varied to obtain the probability of false alarm, sensing time and maximum throughput.

The addition of more cognitive radios in the cooperative sensing scheme increases general communication overhead and energy consumption. It is, therefore, necessary to investigate the optimum number of cognitive radios in a communication network to ensure high-quality performance in terms of resource conservation.

## 2  System Model

### 2.1  Theoretical Background

Energy detection is used energy characteristics of the channel to detect the presence of a primary (licensed) user within a particular communication channel [22]. In energy detection, the energy detected in the sensed channel is measured and compared with a preset threshold to identify the operation of a primary user (PU) [6].

The energy detection process involves receiving a signal $x(t)$, which is filtered by a bandpass filter (BPF). This is within sensed frequency channel limits. This signal detected is then squared with a square-law device. The band pass filter reduces the noise bandwidth. The remaining noise, therefore, has a flat spectral density, which is band-limited. The squaring device prepares the signal for the integrator, which gives the energy value of the signal at interval $T$. Afterward, the output signal from the integrator, $Y$, is compared using a threshold device to decide to ascertain the availability of the PU in the channel. The decision regarding the usage of the band will be made by comparing the detection statistic to a threshold [23].

The mathematical model for energy detection is given by the following two hypotheses [7]:

$$H_0 : \text{PU absent}$$

$$y(n) = u(n) \quad n = 1, 2, \ldots N \tag{1}$$

$$H_1 : \text{PU present}$$

$$y(n) = s(n) + u(n) \quad n = 1, 2, \ldots N \tag{2}$$

where $u(n)$ is noise and $s(n)$ is the PU's signal.

Energy detector performs well in spectrum sensing if the noise variance is known. This is needed to set the threshold which is needed to decide if the channel is occupied or not [24]. The challenge with energy-detection based spectrum sensing is its inability to detect the PU at low SNRs accurately. The detection accuracy further worsens when the noise characteristics cannot be defined due to varying noise uncertainties [25, 26].

This study is aimed at reducing interference, which may occur in energy-detection based cognitive radio through the inclusion of supervised machine learning. It is expected that the cognitive radio system (CRS) will be able to learn the patterns in the unknown noise characteristics through a clustering algorithm. Specific properties of the PU were used as training data in a supervised learning technique to serve a feature detection algorithm in the CRS. This process is intended to improve the detection accuracy of the energy detector in scenarios when the SNR falls to the SNR wall level.

Equation (3) shows the normalized test (decision) statistic for the detector and this was developed based on [27] as:

$$T' = \left(\frac{1}{N_{02}}\right) \int_0^T y^2(t)dt \tag{3}$$

where:

$T'$ = test statistic in during sensing session
$y$ = received signal input
$T$ = sampling instant
$N_{02}$ = two-sided noise power density spectrum

If the test statistics exceeds a fixed decision threshold, then it results in $H_1$ hypothesis. However, when the test statistics is less than the decision threshold, then $H_0$ hypothesis occurs.

As shown in [13], $\lambda$ is the decision threshold which in the number of samples $N \gg 1$, can be expressed as a Gaussian distribution:

$$\lambda = \sqrt{\frac{2}{NQ^{-1}}}\left(P_{fa}^{CED} + 1\right) \tag{4}$$

where:

$$P_{fa}^{CED} = Q\left(\frac{\lambda - 1}{\sqrt{\frac{2}{N}}}\right) \tag{5}$$

$$P_d^{CED} = Q\left(\frac{\lambda - (1+\gamma)}{\sqrt{(\frac{2}{N})(1+\gamma)^2}}\right) \tag{6}$$

$$\gamma = \frac{\sigma_s^2}{\sigma_w^2} \tag{7}$$

$\sigma_s^2$ is the received average PU signal power
$\sigma_w^2$ is the noise variance.

## 2.2  Optimization Model

The operating characteristics in the network can be assessed in frames $(N)$. The energy test statistic $(Y_{p|s|x,i}^{\alpha})$ at the $i^{th}$ frame of the user's transmission operations was extracted as input data [28]. Similarly, $Y_{p|s|x,i}^{\beta}$ and $Y_{p|s|x,i}^{\gamma}$ was extracted at specific points in the channel and receiver respectively.

Energy test statistics for the primary user $(Y_{p,i})$ is given as [28]:

$$Y_{p,i} \in (Y_{p,i}^{\alpha}, Y_{p,i}^{\beta}, Y_{p,i}^{\gamma}) \tag{8}$$

Energy test statistics for a secondary user $(Y_{s,i})$ is given as [28]:

$$Y_{s,i} \in (Y_{s,i}^{\alpha}, Y_{s,i}^{\beta}, Y_{s,i}^{\gamma}) \tag{9}$$

Energy test statistics for an interfering user $(Y_{x,i})$ is given as [28]:

$$Y_{x,i} \in (Y_{x,i}^{\alpha}, Y_{x,i}^{\beta}, Y_{x,i}^{\gamma}) \tag{10}$$

The labels identifying these input data in specific frames as primary user ($U_p$), secondary user ($U_s$) or interfering user ($U_x$) based on their respective energy test statistics are represented as decisions ($d_i$).

Genetic Algorithm (GA) and Simulated Annealing (SA) were used to optimize the number of cognitive radios considering constraints such as the spatial distribution of the cognitive radios, sensing time and noise characteristics. This is important to identify the optimal number of cognitive radios that would minimize resource consumption while ensuring interference mitigation in the system. This objective is achieved by the following function [28]:

$$\min_n Y(n) = n_1 Y^{\alpha}_{p,i} + n_2 Y^{\beta}_{p,i} + n_3 Y^{\gamma}_{p,i} + n_4 Y^{\alpha}_{s,i} + \dots$$

$$n_5 Y^{\beta}_{s,i} + n_6 Y^{\gamma}_{s,i} + n_7 Y^{\alpha}_{x,i} + n_8 Y^{\beta}_{x,i} + n_9 Y^{\gamma}_{x,i} \tag{11}$$

Subject to

$$n_1, n_2, n_3, n_4, n_5, n_6, n_7, n_8, n_9 \geq 0 \tag{12}$$

$$n_1 + n_2 + n_3 \geq 3 \tag{13}$$

$$n_4 + n_5 + n_6 \leq 3 \tag{14}$$

$$n_7 + n_8 + n_9 \leq 3 \tag{15}$$

Cooperative sensing aids more accurate detection as shown in [29]. The first constraint in (12) is therefore crucial for the cognitive radios to operate cooperatively to minimize missed detections and false alarms.

The critical nature of primary user detection makes it imperative to include not less than three cognitive radios to cover every zone of operation. This is represented in the second constraint in (13). An assumption made is that three cognitive radios are sufficient to monitor operations of potential secondary users and interfering users. This is based on previous studies [30] where the provision of an extra cognitive radio for monitoring other non-PU operations produces a more accurate sensing outcome.

## 2.3 Simulation

The model was setup using MATLAB Simulink on MATLAB 2017a software. It comprised transmitters and an energy detector based cognitive radio through an additive white Gaussian noise (AWGN) channel. A channel noise variance of $-25$ dB was used. The sensing was carried out at a frequency of 936 MHz at a bandwidth of 200 kHz and a threshold of 0.2. Details of the simulation parameters are presented in Table 1.

To ensure that the sensing accuracy was not affected, the sensitivity of the cognitive radio system was first improved using ML to ensure that the sensing accuracy is not compromised as the optimization is done. Tree algorithms and KNN were selected based on a previous study [31], which showed that these algorithms performed very well. After improving the accuracy of the CRS with ML, GA, and SA were then used to obtain the optimum number of cognitive radios that would minimize energy consumption.

**Table 1.** Simulation parameters.

| Parameters | Value |
|---|---|
| SNR | −25 dB |
| $P_f$ | 0.05 |
| Operating frequency | 936 MHz |
| Observation time | $2 \times 10^{-4}$ |
| Variance of the noise ($\sigma^2 n$) | $1 \times 10^{-12}$ |
| Threshold | 0.2 |
| Operating power | 40 mW |
| Bandwidth | 200 kHz |
| Population type | Double vector |
| Population size | 200 |
| Scaling function | Rank |
| Selection function | Stochastic uniform |
| Crossover fraction | 0.8 |
| Crossover function | Constraint dependent |
| Generations | 900 |
| Annealing function | Fast annealing |
| Temperature update function | Exponential temperature update |
| Acceptance probability function | Simulated annealing acceptance |

## 3   Results and Discussion

Classification algorithms such as Complex Tree, Fine KNN, Weighted KNN, Cubic KNN, and Medium KNN used to train the data. Table 2 shows the accuracy of the different classifiers considered at −25 dB. The improvement in the specificity of cognitive radio detection to minimize interference is presented in the receiver operating characteristics (Fig. 1). The result compares the accuracy of specificity for the ML-improved cognitive radio and the conventional version at −25 dB.

The ML-improved cognitive radio gave a higher detection accuracy when compared with the conventional cognitive radio. Complex tree classifier gave the highest detection accuracy for PU at >99%. The detection accuracy for the SU and IU were 85% and 87% respectively, and overall detection accuracy of 90.5%. Nine CRs were used for sensing for each category of users (PU, SU, IU) with the assumption that they will produce more accurate results.

Figure 2, presents the comparison of the outcome of using SA and GA to minimize the number of cognitive radios as against the conventional cooperative sensing where all the cognitive radios are utilized. For PU, GA and SA used six and four cognitive radios respectively for sensing at various locations. Thus, using fewer resources without compromising the detection accuracy. The results also reveal that for GA, five cognitive

**Table 2.** Accuracy of classifiers at −25 dB

| CM | PU | | SU | | IU | | OA (%) |
|---|---|---|---|---|---|---|---|
| | TPR (%) | FNR (%) | TPR (%) | FNR (%) | TPR (%) | FNR (%) | |
| WKNN | 93 | 7 | 89 | 11 | 91 | 9 | 91.1 |
| FKNN | 91 | 9 | 89 | 11 | 92 | 8 | 90.8 |
| MKNN | 95 | 5 | 88 | 12 | 89 | 11 | 90.7 |
| CKNN | 95 | 5 | 88 | 12 | 89 | 11 | 90.4 |
| CTree | >99 | <1 | 85 | 15 | 87 | 13 | 90.5 |

Key
CM: Classification methods
WKNN: Weighted KNN
FKNN: Fine KNN
MKNN: Medium KN
CKNN: Cubic KNN
CTree: Complex tree
TPR: True positive rate
FNR: False negative rate

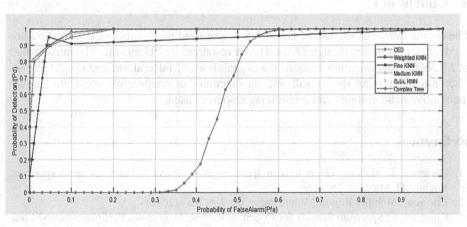

**Fig. 1.** ROC curve comparing five classifiers at −25 dB.

radios each are sufficient to sensing the activities of the SU and the IU. However, SA used 7 and six cognitive radios to detect the events of the SU and the IU, respectively. In terms of minimization of resource utilization, GA gave an overall reduction in energy of 40.74% while SA had a 37.03% reduction.

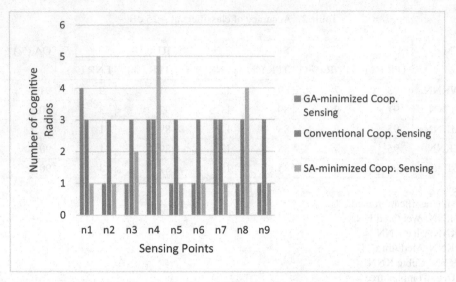

**Fig. 2.** Number of CRs per sensing point.

## 4   Conclusion

The results show that the use of GA and SA optimized the number of cognitive radios used per sensing period though SA used lesser resources for PU sensing. The introduction of ML before optimization reduced the probabilities of false alarm and misdetections. Therefore, improving the overall sensing outcomes of cooperative spectrum sensing incognitive radio systems and minimizing resources used.

## References

1. Chiwewe, T., Hancke, G.: A look at spectrum management policies for radio spectrum. EngineerIT, March 2015
2. Ericsson, "Mobility Report," White Paper, pp. 7–8, May (2016)
3. Chaudari, S.: Spectrum Sensing for Cognitive Radios: Algorithms, Performance, and Limitations, Aalto University School of Electrical Engineering, Department of Signal Processing and Acoustics (2012)
4. Haykin, S., Thomson, D.J., Reed, J.H.: Spectrum sensing for cognitive radio. Proc. IEEE **97**(5), 849–877 (2009)
5. Reddy, G.S.A.K., Raju, U.G., Aravind, P., Sushma, D.: Intelligent wireless communication system of cognitive radio. **5**, 78–82 (2013)
6. Mmary, C.: Cognitive Radio for Broadband Access in Rural Africa and other Developing Countries. MSc. Thesis, University of York, UK, December 2011
7. Sudeep, S., Nirajan, K.: Energy detection based techniques for spectrum sensing in cognitive radio over different fading channels. J. Selected Areas Telecommun. **4**(2), 15–22 (2014)
8. Marcus, M., Burtle, J., Mcneil, N., Lahjouji, A., McNeil, N.: Report of the unlicensed devices and experimental licenses working group. In: FCC, Spectrum Policy Task Force, pp. 1–24 (2002)

9. Mittal, R., Garg, E.D.: A review on spectrum sensing techniques. Int. J. Adv. Res. Comput. Sci. Softw. Eng. **5**(5), 1187–1192 (2015)
10. Lakshmi, M., Saravanan, R., Muthaiah, R.: Energy detection based spectrum sensing for cognitive. Int. J. Eng. Technol. (IJET) **5**(2), 963–967 (2013)
11. Verma, P.K., Taluja, S., Lal Dua, R.: Performance analysis of energy detection, matched filter detection & cyclostationary feature detection spectrum sensing techniques. Int. J. Comput. Eng. Res. **2**(5), 2250–3005 (2012)
12. Ustok, R.F.: Spectrum sensing techniques for cognitive radio systems with multiple antennas. Izmir Institute of Technology (2010)
13. López-Benítez, M., Casadevall, F.: Improved energy detection spectrum sensing for cognitive radio. IET Commun. **6**(8), 785–796 (2012)
14. Kanti, M., Barma, D., Singh, H., Roy, S., Sen, S.K.: Augmented spectrum sensing in cognitive radio networks. IJCSN Int. J. Comput. Sci. Netw. **4**(6) (2015)
15. Zhu, J., Song, Y., Jiang, D., Song, H.: Multi-armed bandit channel access scheme with cognitive radio technology in wireless sensor networks for the Internet of Things. IEEE Access **4**, 4609–4617 (2016)
16. Jain, M., Kumar, V., Gangopadhyay, R., Debnath, S.: Cognitive radio oriented wireless networks. In: CROWNCOM 2015, LNICST, vol. 156, pp. 225–234 (2015)
17. Raj, V., Dias, I., Tholeti, T., Kalyani, S.: Spectrum access in cognitive radio using a two-stage reinforcement learning approach. IEEE J. Sel. Top. Sign. Proces. **12**(1), 20–34 (2018)
18. Kumar, A., Thakur, P., Pandit, S., Singh, G.: Analysis of optimal threshold selection for spectrum sensing in a cognitive radio network: an energy detection approach. Wireless Netw. **25**(7), 3917–3931 (2019). https://doi.org/10.1007/s11276-018-01927-y
19. Wang, H., Jiang, F., Zhou, M.: Cognitive radio power allocation algorithm based on improved particle swarm optimization. In: IEEE International Conference on Communication Systems (ICCS), pp. 354–359 (2018)
20. Elhachmi, J., Guennoun, Z.: Cognitive radio spectrum allocation using genetic algorithm. EURASIP J. Wireless Commun. Netw. **2016**(1), 1–11 (2016). https://doi.org/10.1186/s13638-016-0620-6
21. Kochar, S., Garg, R.: Spectrum sensing for cognitive radio using genetic algorithm. Int. J. Online Biomed. Eng. **14**(9), 190–199 (2019)
22. E. Union, West African Common Market Project: Harmonization of Policies (2008)
23. Nirajan, K., Sudeep, S., Suman, S., Lamichhane, B.: Performance comparison of energy detection based spectrum sensing for cognitive radio networks. Int. Refer. J. Eng. Sci. (IRJES) ISSN, **49**(8) 2319–183 (2015)
24. Axell, E., Leus, G., Larsson, E.G., Poor, H.V.: Spectrum sensing for cognitive radio: state-of-the-art and recent advances. IEEE Signal Process. Mag. **29**(3), 101–116 (2012)
25. Hoven, N., Tandra, R., Sahai, A.: Some fundamental limits on cognitive radio. Wireless Foundations EECS, University of California at Berkeley (2005)
26. Axell, E., Larsson, E.G.: Optimal and sub-optimal spectrum sensing of OFDM signals in known and unknown noise variance optimal and sub-optimal spectrum sensing of OFDM signals in known and unknown noise variance. IEEE J. Sel. Areas Commun. **29**(2), 290–304 (2011)
27. Urkowitz, H.: Energy detection of unknown deterministic signals. Proc. IEEE **55**(4), 523–531 (1967)
28. Fajemilehin, T., Yahya, A., Langat, K., Opadiji, J.: Optimizing cognitive radio deployment in cooperative sensing for interference mitigation. In: BIUST Research and Innovation Symposium 2019 (RDAIS 2019), vol. 2019, no. June, pp. 76–81 (2019)
29. Fajemilehin, T.O., Olatunji, S.A., Opadiji, J.F.: Improved energy detection algorithm for cognitive radios in cooperative spectrum sensing. Int. J. Inf. Process. Commun. (IJIPC) **7**(1), 148–163 (2019)

30. Opadiji, J.F., Olatunji, S.A., Fajemilehin, T.O.: On energy detection of cognitive radios in cooperative spectrum sensing. In: URSI-NG Conference Proceedings, pp. 29–36 (2015)
31. Mikaeil, A.M.: Machine learning approaches for spectrum management in cognitive radio networks. In: Farhadi, H. (ed.) Machine Learning - Advanced Techniques and Emerging Applications, pp. 117–140. IntechOpen, Rijeka (2018)

# Real-Time Sickle Cell Anemia Diagnosis Based Hardware Accelerator

Mohammed A. Fadhel[1,4(✉)], Omran Al-Shamma[1], Laith Alzubaidi[1,3], and Sameer Razzaq Oleiwi[2]

[1] University of Information Technology and Communications, Baghdad, Iraq
{Mohammed.a.fadhel,o.al_shamma}@uoitc.edu.iq
[2] College of Nursing, Muthanna University, Muthanna, Iraq
samirrazak@mu.edu.iq
[3] Faculty of Science and Engineering, Queensland University of Technology, Brisbane, Australia
laith.alzubaidi@hdr.qut.edu.au
[4] College of Computer Science and Information Technology, University of Sumer, Thi Qar, Iraq

**Abstract.** Sickle cell anemia (SCA) is a blood disease, which causes distortion in the shape of Red Blood Cells (RBCs) and becomes like a crescent. Traditional methodologies of classifying and counting RBCs that have been used by medical analysts are time-consuming, as well as, cost-effective. In addition, it is possible to make errors throughout the classifying and counting stages. We overcome these limitations by proposing a novel convolutional neural network model that classifies the RBCs into three categories: Normal, Abnormal (sickle shape) and other blood content. We enhanced the model's efficiency using a hardware accelerator called FPGA (Altera DE2 Cyclone II) to take advantage of its parallelism features. We have evaluated the model's efficiency with different platforms to show the differentiation in terms of time execution and power computation. Based on our knowledge, the results show that our proposed model has achieved the best accuracy (87.15%) and has high efficiency for real-time diagnosis.

**Keywords:** CNN · SCA · Accuracy · Efficiency · FPGA

## 1 Introduction

One of the inherited blood disorders is sickle cell disease. Hemoglobin is the main part of the red cell. It is responsible for carrying the oxygen to the whole body cells through the blood. The Flexibility and circularity are two RBC characteristics that allow them to travel inside the small blood vessels. In general, red cell age is around 120 days. On the other hand, sickle blood cells have a crescent shape, which makes it easy to break apart. Ultimately, Anemia disease is the result, as these cells stay alive only ten to twenty days [1]. The irregularity in the hemoglobin gene is the main reason for sickle cell disease. As long as the oxygen is free from the hemoglobin of the sickle cell, it joins collectively to form lengthy bars, which in turn, defect and adjust the red blood cell shape [2].

The main symptoms of sickle cell disease are the development of awful pain in the abdomen, legs, arms, back, and chest. It can be anywhere in the body. For instance,

© Springer Nature Switzerland AG 2020
A. M. Al-Bakry et al. (Eds.): NTICT 2020, CCIS 1183, pp. 189–199, 2020.
https://doi.org/10.1007/978-3-030-55340-1_14

dangerous illness with difficulty breathing, fever, and chest pain is the result of sickle cell disease in the lungs. Moreover, symptoms and dangerousness are greatly different from one person to another, yet inside the same family. In general, symptoms of sickle cell disaster involve serious infection, blood flow blockage in the liver or spleen, strokes, difficulty in breathing and chest pain, anemia, and awful pain. From the diagnostic side view, examining a blood sample with a microscope, the sickle cells can easily recognize. This blood test is known as hemoglobin electrophoresis. It determines the quantity of the abnormal sickle hemoglobin [3]. Hence, the person is classified either as having sickle cell disease or as a carrier, based on the quantity of the sickle hemoglobin. In addition, Amniocentesis or chorionic villus sampling is used for testing the DNA of the fetal cells, to show the possibility of the prenatal examining for sickle cell disease. Reducing the hazard of more complications and critical infections is achieved by early detection and treatment [4]. Since blood health is extremely significant in human health, early disease recognition is required for avoiding the worsening of the disease and death [5].

However, in scientific research, image-processing technology becomes a fundamental and efficient tool, especially in the biomedical engineering field [6]. Therefore, looking for more effective and accurate methodologies to diagnose sickle cell disease is a necessary and important approach. This study demonstrates the image processing technology as a useful method for sickle disease recognition. It implies of using a novel model for classifying the RBCs with a hardware accelerator.

## 2  Literature Review

Several studies are released related to the detection and classification fields of sickle blood cells. Image segmentation and form factor are used for detecting the abnormality of the blood cells [7]. In this study, the circular shape of the blood cell is examined by the form factor for counting the abnormal cells. Other work assigned a threshold for each abnormal blood cell type [8]. However, the image processing means executing some processes on the image to improve it or to carry out specific information that is outlined with predefined principles [9]. It is one of the computer science branches. More specifically, image processing is a basic technique for blood cell recognition. Image acquisition is achieved by interfacing the digital microscope to the computer. The output image from the microscope is a digital image and its resolution is based on the used microscope type. Next, image enhancement is taking place in order to improve the image quality by improving the image details [2]. Image segmentation is the next step. The digital image is partitioned into several segments. Each segment is actually a set of pixels. It is a so-called super pixel. The main objective of this step is to make the image simpler and/or to modify its description into extra simple and significant meaning for analyzing. In general, image segmentation is utilized for locating boundaries and objects (curves, lines, etc.) inside the image. More specifically, it is a label assigning process for each pixel in the image. Hence, similar labels allocate the same characteristic [10].

In the final step, the detection of abnormal blood cells is achieved by using a form factor threshold, which is constant for various abnormal blood cells [1]. Starting by inputting the image and converts it into a black-white scale (so-called grayscale) image. Next, detecting the abnormal cells based on the edge detecting method. It is an image segmentation technique. It finds out the edge or line presence in the image and delineates it in a suitable way. The key objective of this step (edge detection) is for shortening the data, and thus, minimizing the data amount for the next processing [9]. Labeling and form factor computing step is followed by the edge detection step [12].

The normal cell is identified if the form factor is in the range 0.6–1, while the abnormal cell is identified if the form factor is below 0.6. Hence, the sickle blood cell is identified if the form factor is below 0.5 [13]. The main benefits of this technique are its speed, simplicity, and detecting the cell abnormality by calculating the cell circularity. The main drawback is its inability to recognize two attached normal cells and counted them as a single abnormal cell. Circle Hough transform is used to detect the sickle cells [1]. Recently, deep neural network interferes with medical applications to obtain high accuracy of classification [3, 11]. Traditional methodologies of classifying and counting RBCs that have been used by medical analysts are time-consuming, as well as, cost-effective. In contrast, our work presents a novel model of RBCs classification that built-in a hardware accelerator. The model enhanced the classification accuracy and achieved high efficiency for real-time diagnosis.

## 3  Methodology

The methodology consists of three main stages: data-set acquisition, training patches, and structure of Convolutional Neural Networks:

### 3.1  Data-Set

Initially, the first step is preparing a data-set of different color RBC images, which involve normal and abnormal types. We collected 202 images from various standardized websites like references [14, 15]. We collected blood smear samples based on Wadsworth center data. The samples have various colors, sizes, and shapes. They are very noisy, and in fact, make the classification task in a challenge. Figure 1 shows one sample of input data-set. All images are cropped to the size of 28 * 28 pixels and are separated into 15600 patches of normal, abnormal and suspicious cases. The expert physician put a label on all images with their classes: normal, abnormal (crescent shape) and other blood contents. The eighty percent patches are used for the training stage and the remaining twenty percent patches are used for the testing stage.

### 3.2  Augmentation Process for Training Patches

Deep convolutional neural network has a large number of parameters. The learning algorithm is required for tuning the primarily weights, which are linked to the convolutional layers.

In addition, collecting many medical images is a cost-effective and challenging job. We have implemented patches to improve our model's performance. We used multiple techniques of image processing such as rotation, flipping, and improved color contrast using different color spaces techniques to increase training data. Rotation was done with angles of 45, 90, 180, 220, and 270. Next, flipping is achieved by three ways (horizontal, vertical and horizontal vertical). The patches are repeated for offline training ten times according to the data augmentation principles [16].

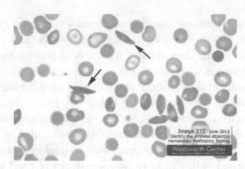

**Fig. 1.** The input image sample (note: the arrows mark on to SCA).

### 3.3 Principles of Convolutional Neural Networks (CNNs)

Convolutional Neural Networks (CNNs) play an important role in image classification. They can be applied to healthcare, security, Facebook's tagging, and Auto-driving vehicles. CNNs compose of three key-layers [17], as illustrated in the following, and shown in Fig. 2.

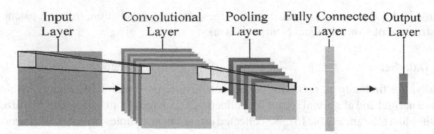

**Fig. 2.** Layers of Convolutional Neural Networks

**The Convolutional Layer.** It convolves the past layer (input layer) output with a filter set for training and learning [18, 19]. The weights are distributed to the convolution filter. Each part of the filter moved row-by-row across the input volume, generating an activation map with two dimensions (height and width) of that filter. The depth of filters and input are the same.

However, three parameters can be controlled the output size, which are: Stride, Zero-padding, as well as, Depth. Depth is described as a group of filters that convolve to the input image. Each filtertests features like blobs, corners, edges, etc. Stride is a filter group of pixels moves sliding over the input image. Lastly, Zero-padding means filling zeros on the image boundaries for saving its size.

**Pooling Layer.** It is located in the middle of the network, after the convolutional layer. The task of the pooling layer is to condense the convolved feature maps. There are two popular operators, average and max pooling. These operators find the average or maximum value over the spatial block. The common pooling filter size is 2 * 2 square with two strides.

**Fully Connected Layer (FC).** It is typically the last layer of the CNN. As a simple Neural Networks, neurons are used for connecting the whole activations in the past layer. The main function of the FC is to achieve the classification task.

### 3.4 The Proposed Model Architecture

Our proposed network is designed based on the idea of a multi-branch network. It has two branches of parallel convolutional layers. This type of architectures is helpful in a way of extracting different levels of features. It is extremely valuable for back propagation since the error can back propagate from various paths. In addition, it has the prompt benefit of enlarging its width without drastically enlarging its computational cost, which can lead to enhance the performance of the network. We present a CNN architecture of 21 layers with five convolutional layers (CONV1, CONV2, CONV3, CONV4, and CONV5) as shown in Fig. 3. The input size of 28 * 28 * 3 is fed to the first convolutional layer, which extracts features such as edges, color, and shape, and helps to reduce the

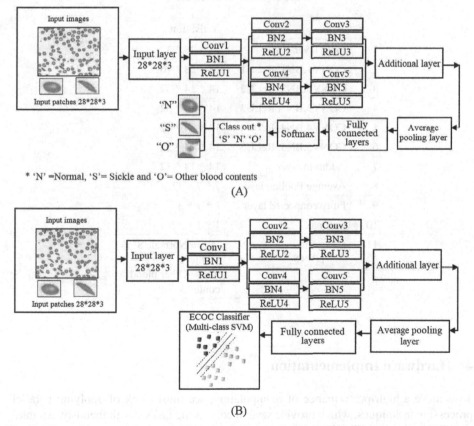

**Fig. 3.** The architecture of suggested Model A) Our Model B) ECOC classifier mixed with our model

input image size. The output of the first convolutional layer (CONV1) is fed to CONV4 and CONV2. The output of CONV2 is passed to CONV3, while the output of CONV4 is passed to CONV5. Afterward, the outputs of CONV5 and CONV3 are added together in an Additional layer. Latter, it feeds to an average-pooling layer for reducing the number of training parameters. Each convolutional layer is followed by Batch Normalization (BN) and Rectified Liner Unit (ReLU) which is defined as $f(x) = \max(0, x)$. ReLU helps to expedite the training process. Lastly, a fully connected layer converts the features as one-vector features, and then it passes to softmax function as illustrated in Fig. 3A and Table 1. For the comparison purposes, we have trained ECOC (Error-Correcting Output Codes) classifier to classify three classes. The ECOC classifier is applied to reduce a problem of multiclass to a problem of binary classification. Conversely, ECOC classifier showed well-behaved results in various tasks [20]. Therefore, the extracted features of our network are employed for training the ECOC classifier, as displayed in Fig. 3B. ECOC classifier is based on the reduction idea to multi-binary classifiers like support vector machines (SVMs).

**Table 1.** Model architecture

|    | Name                       | Activation                                          |
|----|----------------------------|-----------------------------------------------------|
| 1  | Input layer                | 28 * 28 * 3                                         |
| 2  | CONV1.BN1, ReLU1           | 28 * 28 * 16                                        |
| 3  | CONV1.BN2, ReLU2           | 14 * 14 * 32                                        |
| 4  | CONV3.BN3, ReLU3           | 14 * 14 * 32                                        |
| 5  | CONV4.BN4, ReLU4           | 14 * 14 * 32                                        |
| 6  | CONV5.BN5, ReLU5           | 14 * 14 * 32                                        |
| 7  | Addition layer             | 14 * 14 * 32                                        |
| 8  | Average Pooling layer      | 7 * 7 * 32                                          |
| 9  | Fully connected layer      | 1 * 1 * 3                                           |
| 10 | Softmax                    | 1 * 1 * 3                                           |
| 11 | Class out 'S' 'N' 'O'      | 'N' = Normal 'S' = Sickle 'O' = Other blood contents |

## 4   Hardware Implementation

To achieve a high performance of manipulation, we must think of applying parallel processing techniques, which provide several computing tasks simultaneously. An integrated circuit called Field-programmable gate array (FPGA) is very useful, since it can be possible to benefit their parallel nature in terms of the number of gates applied for

**Table 2.** Summary of models

| No. | Module | Details |
|---|---|---|
| 1. | DE2 TV.v | The main file calls all subcode parts |
| 2. | TD Detect.v | Check if the video input is valid |
| 3. | ITU 656 Decoder.v | Switch the video standard from ITU 656 to YUV 4:2:2 |
| 4. | DIV.v | This module applied to decrease the horizontal pixels per frame from 720 to 640 |
| 5. | Sdram Control 4port.v | A temporary memory used to mix the odd and even frame |
| 6. | YUV422 to 444.v | Switch the video standard from 4:2:2 to 4:4:4 |
| 7. | YCbCr2RGB.v | Switch the video word length from 8 to 10 bits to compatible with DAC signal |
| 8. | VGA Ctrl.v | A module used to control the output of the VGA screen |
| 9. | Line Buffer.v | A line frame buffer delay is applied for deinterlacing the image |
| 10. | I2C AV Config.v | This module is designed to control the video decode |
| 11. | Our model.v | This model contains our proposed model |
| 12. | Our model+ECOC.v | This model contains our proposed model with ECOC.v |

specific processes. FPGA was used for speeding up the process of classification in real time [24]. Firstly, the data are read from the digital microscope camera through an AV Cable. It has an ITU-656 format, as illustrated in Fig. 4, with YUV 4:2:2 standards, which is so-called YCbCr. Note that Y referred to luma, Cb and Cr referred to blue and red differences separately [21].

After the input became in YCbCr format, the downsampling is touched only the horizontal axis from 720 to 640 pixels because the vertical axis is fixed by 480 pixels. Then, these converted data are supplied to a SDRAM FIFO, which acts as a frame buffer. The result of FIFO is transformed from YUV 4:2:2 to 4:4:4 standard. Next, the new output is converted to 10-bit RGB standard [22, 23].

The next stage is the processing of RGB data through our proposed network, as described in the following section. The network output is delivered, through the VGA controller, to the VGA screen for displaying. All the above description is illustrated in Fig. 5 and the functionality description of different models, shown in the figure, are summarized in Table 2. The FPGA design is written in Verilog language (.v) based on MATLAB code from the reference [3] with few modifications to enhane its accuracy.

The available software that compatible with FPGA board (Altera Cyclone® II 2C35 device) is called Quartus II 11.1 web edition software (32-Bit). The hardware design in Fig. 5 despite the flow of video signals through FPGA hardware starting from TV decoder and reaching video DAC 7123 via our model stage.

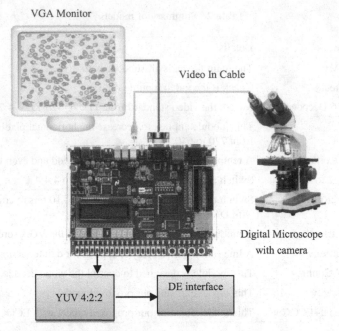

Fig. 4.  The hardware architecture of SCA diagnosis

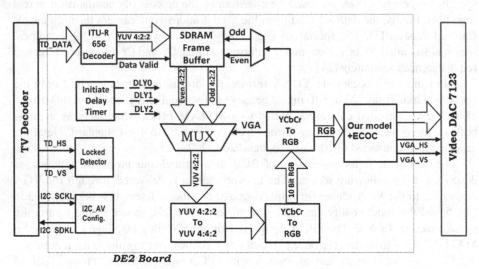

**Fig. 5.**  Sequence diagram for SCA diagnosis used Altera DE2 board

## 5  Results and Discussions

During the CNN development, the ModelSim v10.0c is employed for simulating the architecture behavior. This network is tested on Altera DE2 with chip Cyclone II

EP2C35F672C6 in order to make sure that the hardware design is synthesizable and do correctly on the FPGA. Initially, the diagnostic accuracy function of the CNN is calculated as the proportion between the number of patches, that have been correctly classified, and the total number of patches. Our model is tested for different images with different shape, color and overlapped cells. In addition, our model is able to solve all these challenges. The shape feature is an important feature for distinguishing between classes.

As shown in Table 3, we evaluated our model in terms of patch wise classification by dividing the input images into small patches of 28 * 28 pixels. The highest probability is assigned to match classes. The accuracy of our proposed model is 85.74%, and when adding ECOC to our model, the accuracy is enhanced to 87.15%.

**Table 3.** The percentage rate of classification and time of execution

|   | Models | Patch-wise% | Time of execution in msec used FPGA |
|---|--------|-------------|-------------------------------------|
| 1 | Our model | 85.74 | 48.0818 |
| 2 | Our model+ECOC | 87.15 | 63.7408 |

Figure 6 illustrates different samples of the tested patches. Our model has a well-behaved performance of the RBC classification. The probabilities reflect the level of the SCA disease. For instance, the case S = 97.17%, N = 1%, and O = 1.83%, means it is in the dangerous level and the specialist physician must take the necessary actions.

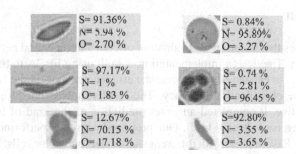

S= 91.36%
N= 5.94 %
O= 2.70 %

S= 0.84%
N= 95.89%
O= 3.27 %

S= 97.17%
N= 1 %
O= 1.83 %

S= 0.74 %
N= 2.81 %
O= 96.45 %

S= 12.67%
N= 70.15 %
O= 17.18 %

S=92.80%
N= 3.55 %
O= 3.65 %

**Fig. 6.** Some of the tested patches of our suggested model with ECOC classifier

As we explained previously, our system is designed using the FPGA system to overcome the delay of the training and testing tasks inside the CNN.

On the other hand, we calculated the reserved size inside the FPGA in terms of the logic elements, registers, memory, and embedded multiplier, as clarified in Table 4. Our model with ECOC is has larger hardware architecture due to the additional part of the ECOC classifier.

In this work, we calculated the various layout options in the FPGA, but it is still significant to compare with another platform such as the CPU. Thus, Intel® Core™ i7 CPU 650 2.60 GHz is used. Table 5 lists the performance comparison between the FPGA and the CPU (250 MHz and 2600 MHz respectively), with respect to the time execution and the power consumption.

**Table 4.** Flow summary of the compilation report

|   | Models | Total logic elements out of 33,216 | Total registers | Total memory out of 483,840 | Embedded multiplier out of 70 |
|---|---|---|---|---|---|
| 1 | Our model | 20,678 | 7833 | 53,184 | 18 |
| 2 | Our model+ECOC | 23,417 | 7989 | 63,213 | 20 |

**Table 5.** Performance comparison of different platforms

|   | CPU | FPGA |
|---|---|---|
| Time (sec) | 4.332 | 0.0637 |
| Power (mW) | 31000 | 149.48 |

Finally, we conclude that the FPGA design has lower time calculation than the CPU, taking into account that, the MATLAB language has not get the highest performance language among the available languages. In addition, the FPGA has lower power consumption due to the minimum logic components used.

## 6 Conclusion

Our work provides the design space evaluation of convolution neural network for sickle cell classification. The design implementation is built using FPGA to be used for real-time purposes and to enhance the performance efficiency. We used the ECOC classifier for increasing the classification accuracy. The results showed our proposed model is robust and effective. It completed an exact prediction of the kind of blood cells that reflected the dangerous disease level. Our proposed approach performed very well in classifying noisy RBC images with different shape, size, and color cells. In addition, the model is applied with different platforms (CPU and FPGA) for evaluating the execution time and the power consumption. Finally, based on our knowledge, our model has the best accuracy and a high efficiency of sickle cells classification.

## References

1. Fadhel, M.A, Humaidi, A.J., Oleiwi, S.R.: Image processing-based diagnosis of sickle cell anemia in erythrocytes. In: 2017 Annual Conference on New Trends in Information & Communications Technology Applications (NTICT), pp. 203–207. IEEE (2017)
2. Alzubaidi, L., Fadhel, M.A., Al-Shamma, O., Zhang, J.: Robust and efficient approach to diagnose sickle cell anemia in blood. In: Abraham, A., Cherukuri, A.K., Melin, P., Gandhi, N. (eds.) ISDA 2018 2018. AISC, vol. 940, pp. 560–570. Springer, Cham (2020). https://doi.org/10.1007/978-3-030-16657-1_52

3. Alzubaidi, L., Al-Shamma, O., Fadhel, M.A., Farhan, L., Zhang, J.: Classification of red blood cells in sickle cell anemia using deep convolutional neural network. In: Abraham, A., Cherukuri, A.K., Melin, P., Gandhi, N. (eds.) ISDA 2018 2018. AISC, vol. 940, pp. 550–559. Springer, Cham (2020). https://doi.org/10.1007/978-3-030-16657-1_51
4. Alzubaidi, L., Fadhel, M.A.., Al-Shamma, O., Zhang, J., Duan, Y.: Deep learning models for classification of red blood cells in microscopy images to aid in sickle cell anemia diagnosis. Electronics 9(3), 427 (2020)
5. Fasano, R.M., Booth, G.S., Miles, M., Du, L., Koyama, T., Meier, E.R., et al.: Red blood cell alloimmunization is influenced by recipient inflammatory state at time of transfusion in patients with sickle cell disease. Br. J. Haematol. 168(2), 291–300 (2015)
6. Abubakar, I., Tillmann, T., Banerjee, A.G.: Regional, and national age-sex specific all-cause and cause-specific mortality for 240 causes of death, 1990–2013: a systematic analysis for the Global Burden of Disease Study 2013. Lancet 385(9963), 117–171 (2015)
7. Sreekumar, A., Bhattacharya, A.: Identification of sickle cells from microscopic blood smear image using image processing. Int. J. Emerg. Trends Sci. Technol. 01(05), 783–787 (2014)
8. Darrow, M., Zhang, Y., Cinquin, B.P., Smith, E.A., Boudreau, R., Rochat, R.H., et al.: Visualizing red blood cell sickling and the effects of inhibition of sphingosine kinase 1 using soft X-ray tomography. J. Cell Sci. 129(18), 3511–3517 (2016)
9. Van, E., Samsel, L., Mendelsohn, L., Saiyed, R., Fertrin, K.Y., Brantner, C.A., et al.: Imaging flow cytometry for automated detection of hypoxia-induced erythrocyte shape change in sickle cell disease. Am. J. Hematol. 89(6), 598–603 (2014)
10. Araújo, T., et al.: Classification of breast cancer histology images using convolutional neural networks. PLoS ONE 12(6), e0177544 (2017)
11. Alzubaidi, L., Fadhel, M.A., Oleiwi, S.R. et al.: DFU_QUTNet: diabetic foot ulcer classification using novel deep convolutional neural network. Multimed. Tools Appl. 79, 15655–15677 (2020) https://doi.org/10.1007/s11042-019-07820-w
12. Mahmood, N.H., Che, L.P.: Blood cells extraction using color based segmentation technique. Int. J. Life Sci. Biotechnol. Pharma Res. 2, 2250–3137 (2013)
13. Aruna, N.S., Hariharan, S.: Edge detection of sickle cells in red blood cells. (IJCSIT) Int. J. Comput. Sci. Inf. Technol. 5(3), 4140–4144 (2014)
14. Homepage. https://www.wadsworth.org/. Accessed 12 Sept 2019
15. Homepage. http://sicklecellanaemia.org/. Accessed 15 May 2019
16. LeCun, Y., Bengio, Y., Hinton, G.: Deep learning. Nature 521(7553), 436–444 (2015)
17. Szegedy, C., et al.: Going deeper with convolutions. In: Proceedings of the IEEE Conference on Computer Vision and Pattern Recognition, pp. 1–9 (2015)
18. Alzubaidi, L., Al-Shamma, O., Fadhel, M.A., Farhan, L., Zhang, J., Duan, Y.: Optimizing the performance of breast cancer classification by employing the same domain transfer learning from hybrid deep convolutional neural network model. Electronics 9(3), 445 (2020)
19. Alzubaidi, L., et al.: Towards a better understanding of transfer learning for medical imaging: a case study. Appl. Sci. 10(13), 4523 (2020)
20. Litjens, G., Kooi, T., Bejnordi, B.E., et al.: A survey on deep learning in medical image analysis. Med. Image Anal. 42, 60–88 (2017)
21. Sulaiman, N., Obaid, Z.A., Marhaban, M.H., Hamidon, M.N.: Design and implementation of FPGA-based systems-a review. Aust. J. Basic Appl. Sci. 3, 3575–3596 (2009)
22. Bailey, D.G.: Design for Embedded Image Processing on FPGAs. Wiley, Hoboken (2011)
23. Fadhel M.A., Al-Shamma O., Oleiwi S.R., Taher B.H., Alzubaidi L.: Real-time PCG diagnosis using FPGA. In: Abraham A., Cherukuri A., Melin P., Gandhi N. (eds.) ISDA 2018 2018. AISC, vol 940, pp. 518–529. Springer, Cham. https://doi.org/10.1007/978-3-030-16657-1_48
24. Al-Shamma, O., Fadhel, M.A., Hameed, R.A., Alzubaidi, L., Zhang, J.: Boosting convolutional neural networks performance based on FPGA accelerator. In: Abraham, A., Cherukuri, A.K., Melin, P., Gandhi, N. (eds.) ISDA 2018 2018. AISC, vol. 940, pp. 509–517. Springer, Cham (2020). https://doi.org/10.1007/978-3-030-16657-1_47

# Energy-Efficient Particle Swarm Optimization for Lifetime Coverage Prolongation in Wireless Sensor Networks

Ali Kadhum Idrees[1](✉) ⓘ, Safaa O. Al-Mamory[2] ⓘ, and Raphael Couturier[3] ⓘ

[1] Department of Computer Science, University of Babylon, Babylon, Iraq
ali.idrees@uobabylon.edu.iq
[2] College of Business Informatics, University of Information Technology and Communications, Baghdad, Iraq
salmamory@uoitc.edu.iq
[3] FEMTO-ST Institute/CNRS, The DISC Department, Univ. Bourgogne Franche-Comte, Belfort, France
raphael.couturier@univ-fcomte.fr

**Abstract.** Preserving sufficient coverage and prolonging the network lifetime as much as possible has become one of the most critical issues in Wireless Sensor Networks (WSNs). In this article, a protocol called Energy-efficient Particle Swarm Optimization for Lifetime Coverage Prolongation (EPSOLCOP) is proposed to maintain the coverage and enhance the WSN lifetime. The target sensing field is virtually partitioned into smaller subfields. EPSOLCOP protocol is distributed on the sensor nodes of each subfield in the WSN. The lifetime of each subfield is divided into discovery and activity rounds of the same length. After a neighbor discovery, each round consists of three phases: cluster head election, sensor activity scheduling-based particle Swarm Optimization (PSO), and monitoring phase. The cluster head executes the PSO so as to produce the best representative set of sensor nodes which are responsible for covering the subfield in the next phase. Each set is produced to ensure the coverage at a low energy cost, allowing to optimize the WSN lifetime. In comparison with some existing protocols, simulation results done using the discrete event simulator OMNeT++ show that EPSOLCOP protocol is very competitive by achieving a high coverage ratio with reduced energy consumption.

**Keywords:** Wireless Sensor Networks · Coverage · Particle Swarm Optimization · Activity scheduling · Optimization

## 1 Introduction

In the future Internet of Things (IoT), every object around us will be an active node in the Internet that can generate and utilize the information in the network [9, 10]. Wireless Sensor Network (WSN) is the most important element in the IoT architecture due to its large wide applications in several fields such as environmental, habitat monitoring, agriculture, smart home, health-care, military, and so on [2, 10]. WSN consists of a large

© Springer Nature Switzerland AG 2020
A. M. Al-Bakry et al. (Eds.): NTICT 2020, CCIS 1183, pp. 200–218, 2020.
https://doi.org/10.1007/978-3-030-55340-1_15

number of small, inexpensive, and low-energy sensor nodes [1]. The sensor devices cooperate with each other and communicate via multi-hop communication to perform its tasks and transmit the gathered sensed readings to a central node called sink. WSNs are very significant in cyber-physical systems for monitoring the complex physical environment at low cost [3, 21]. In many applications, it is preferred to deploy the sensor devices in a dense way to ensure the full monitoring of the area of interest to increase the lifetime of WSN and improve the quality of surveillance [17]. One of the main challenges in WSNs is how to cover the area of interest as long time as possible while saving the energy of the sensor nodes to extend the lifetime of WSN. The coverage problem represents a fundamental issue in WSNs. Each sensor device has limited resources like such as restrictions in energy, processing, storage, bandwidth, and communication. Since the energy is the most constrained resource inside the sensor node, therefore, turn on all deployed sensor nodes inside the area of interest lead to deplete the battery power of each sensor node and decrease the network lifetime [12, 14]. In order to solve this problem, it is important to schedule the sensor nodes into several sets of sensor nodes schedules which are responsible for covering the sensing field in different rounds so as to extend the WSN lifetime. Each set capable of covering the sensing field for a certain period of time. In this way, some of sensor nodes are turned o˙ to save energy until starting new time period. The principal contributions of this article are summarized as follows.

I.   This article proposes a protocol named Energy-efficient Particle Swarm Optimization for Lifetime Coverage Prolongation (EPSOLCOP) for WSNs. It maintains the coverage and enhances the WSN lifetime. The sensing field is virtually partitioned into smaller fixed geographical clusters. EPSOLCOP protocol is distributed on the sensor nodes of each cluster in the WSN. The lifetime inside each cluster is partitioned into discovery and activity rounds of the same length. For the latter, the round is composed of three phases: cluster head election, sensor activity scheduling-based Particle Swarm Optimization (PSO) and monitoring phase. For the first, the neighbor discovery is achieved. In every subfield, the cluster head runs the PSO so as to provide the best set of sensor nodes which are responsible for covering the subfield of the target sensing field in the current round. This set is produced to ensure the coverage inside the subfield of the sensing field at a low energy cost. EPSOLCOP protocol employs an efficient distributed approach to choose the cluster head every round and inside every subfield in the target sensing field.
II.  Instead of using the optimization solver to solve the optimization model to produce the optimal sensors schedule per round, EPSOLCOP protocol employs a PSO to provide the optimal or near optimal schedule of sensor nodes for each subfield in the target sensing field to ensure the coverage and extend the lifetime of the network. The optimization model is focused on the covering the points of interest that represent the centers of the sensor nodes by a lower number of sensor nodes in each subfield.
III. Several experiments are achieved using OMNeT++ network simulator and the results show the effectiveness of our EPSOLCOP protocol in improving the WSN lifetime whilst preserving the coverage in the area of interest with a suitable ratio for long time as possible. EPSOLCOP protocol is compared with two existing methods: DESK [30] and GAF [32].

The remaining of this paper is organized as follows. Next section shows the related literature in the field. Section 3 introduces Preliminaries about the problem. Section 4 provides the mathematical formulation of the coverage problem. Section 5 is allocated to EPSOLCOP protocol illustration. Section 6 exhibits the experimental results conducted using OMNeT++ simulator. Finally, we provide the conclusions and future works in Sect. 7.

## 2 Related Literature

In the last decade, the lifetime coverage maximization in WSNs has received a great consideration. This topic is investigated and summarized in [28, 31]. Xu et al. [32] proposed a Geography-informed routing algorithm, called GAF to con-serve the energy for ad hoc network. This algorithm utilizes geographic data of location to divide the monitored area of interest into small fixed size cells. Each node inside a cell will be in one of two states: sleep or listen. Only one node remains awake inside the cell to achieve sensing and routing tasks. The work in [6] proposed a protocol to build an Area Dominating Set (ADS). This protocol provides an energy-saving approach for ensuring the connectivity and coverage in WSN. Due to its high time complexity $O(d3)$, the protocol is not scalable for large WSNs. Vu et al. [30] proposed a full distributed energy saving scheduling algorithm for K-Coverage in WSNs, named DESK. DESK works into rounds. It requires only the information of neighbor nodes. In DESK, each node defines its status (Active or Sleep) based on the perimeter coverage model that introduced by Huang et al. [11]. The work in [24] presented a distributed sensor scheduling scheme for ensuring coverage to improve the lifetime of WSNs. A new metric of coverage is employed to provide the suitable sensor in every cluster. In [33], the authors introduced a stochastic k-coverage method. this method schedules the sensor nodes to minimize the active node number based on the stochastic model of sensing. The authors in [25] execute a multiobjective optimization approach, named multi-objective evolutionary algorithm based on decomposition (MOEA/D) to solve the coverage preservation and energy conservation simultaneously in cluster-based WSNs. The work in [26] aims to ensure the coverage for a grid by positioning the required sensor nodes while preserving connectivity between the deployed nodes and a sink node. A branch and bound (B&B) algorithm is developed to solve this problem in optimal way. Mostafaei et al. [22] suggest a new cover set approach based on Imperialist Competitive Algorithm (ICA). ICA chooses the nodes that should be chose in various cover sets to cover all the targets. In [34], the authors proposed two protocols: distributed and centralized k-coverage protocols for WSNs. These protocols employed Coverage Contribution Area to get a lower spatial density of the sensor. They took into account the remaining power of sensor nodes that combined with the spatial density of sensor to ensure the network lifetime extension. In [13], The author proposed genetic algorithm for solving the coverage problem. The results ex-plain that the proposed algorithm can prolong the lifetime of the network. The authors in [16] proposed a distributed protocol for solving the coverage problem for several rounds. this protocol is based on the GLPK optimization solver to produce the solution for the optimization problem. in spite of the protocol can improve the lifetime but it consumes a large power and execution time for when the number of rounds increase during one execution for the algorithm. The work in [13] proposed a scheduling algorithm for providing the best sensor schedule based on real sensor data and using memetic algorithm.

In this article, we propose an Energy-efficient Particle Swarm Optimization protocol for Lifetime Coverage Prolongation (EPSOLCOP) in order to preserve the coverage and extend the lifetime of WSNs. The target sensing field is virtually clustered into smaller subfields. The nodes of each subfield are grouped in one cluster. EPSOLCOP protocol is distributed on the sensor nodes of each cluster in the sensing field. The lifetime of each subfield is divided into discovery and activity rounds of the same length. After a neighbor discovery, each round consists of three phases: cluster head election, sensor activity scheduling-based Particle Swarm Optimization (PSO) and monitoring phase. In each round, one cluster head is selected for each cluster in the sensing field. The cluster head executes the PSO to produce the best sensors set which are capable of covering the subfield in the covering phase of the current round. Each set is produced to guarantee coverage at a low energy cost, allowing to optimize the WSN lifetime.

# 3   Background

In this section, the problem of coverage in Wireless Sensor Network is defined. The binary particle swarm optimization technique will be explained as well as the network model is defined.

## 3.1   The Coverage Problem in WSNs

In a WSN [7], the large number of wireless sensor nodes either placed at specific locations or distributed randomly in a region of interest. In order to operate such a sensor network, it is necessary to determine the coverage of the sensor nodes from defining the problem of the sensing region and coverage and then optimize it. Coverage is important in wireless sensor networks; the ability of a sensor network to monitor and report events is the sole purpose of the system. We can classify into three types:

a)  *Area Coverage*: The main objective is to cover an area. The area coverage requires that the sensing range of working active nodes cover the whole targeting area, which means any point in target area can be covered [5].

b)  *Target Coverage*: The principle goal is to ensure the coverage for a group of targets. Target coverage means that the discrete target points can be covered in any time. The sensing range of working active nodes only monitors a finite number of discrete points in targeting area [23].

c)  *Barrier Coverage*: the main goal is to define the maximal support/breach paths which cross the sensor field. Barrier coverage is expressed as finding one or more routes with starting position and ending position when the targets pass through the area deployed with sensor nodes [20].

In the literature, the several coverage methods are studied can be considered within above demonstrated types of coverage. Our work will concentrate on the area coverage by design and implement a distributed intelligent optimization algorithm by which efficiently select the active nodes that must keep both the coverage and connectivity of the WSN and in the same time improve the WSN lifetime. The aim of the distributed metaheuristic optimization approach is to conserve the energy by switch the redundant sensors to sleeping state.

## 3.2 Binary Particle Swarm Optimization

Particle swarm optimization (PSO) is a stochastic optimization approach based on the population developed by Eberhart and Kennedy [19] in 1995, inspired by flock of bird's social behavior. Its main strength is in its simplicity and fast convergence rates. The main idea of the PSO is inspired by the bird's behavior in which the large group of birds are flying in a random way and seeking the food together. The particles fly in the search space seeking the optimal solution similar to the bird's way in seeking the food. In PSO the individuals are involved by cooperation and competition among the individual themselves through generations. Each particle adjusts its position in the search space ac-cording to its own flying experience and share theirs companions flying experience [27]. In PSO algorithm, each individual can be seen as a particle (candidate solution) in a multidimensional search space, flying with a certain speed and all the particles constitute the swarm. We Suppose that the real value searching space is D-dimensional and P size particles form a swarm. Each particle is looked as a point in the D-dimensional space, and the $\text{Pr}^{\text{th}}$ particle represents a D-dimensional vector $X^{\text{Pr}} = (X^{\text{Pr}}, ..., X_D^{Psize})$. This means that each particle a candidate solution in domain D and will be evaluated by a fitness function f(X). The Prth particle's velocity is the current flying speed and also a D-dimensional vector, is represented as $V^{\text{Pr}} = (V_1^{\text{Pr}}, ..., V_D^{Psize})$. The particle is updated to move towards the better area by the corresponding operators till the best point is found. In the iterative process, each particle's previous best position (local best position) is remembered and denoted $\text{Lb}^{\text{Pr}} = (Lb_1^{\text{Pr}}, ..., Lb_D^{\text{Pr}})$, and the globally best position in the whole swarm is recorded as $\text{Gb} = (Gb_1, ..., Gb_D)$. In each iteration t, the velocity V and the position X are updated using (1) and (2). The update process is iteratively repeated until either an acceptable Gbest is achieved or the maximum number of iterations MAXG is reached.

$$V_d^{Pr}(t+1) = \omega * V_d^{Pr}(t) + c1 * r1 * \left(Lb_d^{Pr}(t) - X_d^{Pr}(t)\right)$$
$$+ c2 * r2 * \left(Gb_d(t) - X_d^{Pr}(t)\right) \tag{1}$$

$$X_d^{Pr}(t+1) = X_d^{Pr}(t) + V_d^{Pr}(t+1) \tag{2}$$

In (1), c1 and c2 are usually in the range of 0 to 2.5; r1 and r2 are in the range {0, 1}. There are three parts in (1): the first part is the speed of particle in the last iteration where in the original PSO there is no inertia weight parameter ($\omega$) but Y. Shi and R. Eberhart [27] added it to the original PSO to improve its performance, the second part represent the cognition part that represents the distance between the current position with local best position of the particle, which indicates the private thinking of the particle, i.e., it can learn on itself. The third part is the "social" part which is the distance between the current position with the global best position, this represents the collaboration between the particles. It can be seen that particles have the ability of self-understanding and learning from other particles, and the particles will go near to their best positions and global best position. The particles in (2) will y to a new position because they got new speeds by (1). As we saw above, the continuous-valued PSO operates in continuous and real number space, it cannot be used to optimize the pure discrete binary combinatorial optimization problem. Kennedy and Eberhart solve this problem by proposing the binary version of the PSO in which the binary PSO algorithm will use the binary coding [18]. In binary PSO the position vector X (candidate solution) was represented by binary values. For update the position X, we need to use the equation in (3) instead of (2).

$$X_d^{Pr}(t+1) = \begin{cases} 0, & if \ r \geq Sig\left(V_d^{Pr}(t+1)\right) \\ 1, & Otherwise. \end{cases} \tag{3}$$

Where r is a random real number in the range [0,1] and $Sig(V_d^{Pr}(t+1))$ is a sigmoid limiting transformation function that defined as in Eq. (4).

$$Sig\left(V_d^{Pr}(t+1)\right) = \frac{1}{1 + e^{-V_d^{Pr}(t+1)}} \tag{4}$$

The local best position for each particle Pr is updated using (5).

$$Lb^{Pr}(t+1) = \begin{cases} Lb^{Pr}(t), & if \ f(X^{Pr}(t+1)) \geq f(Lb^{Pr}(t)) \\ X^{Pr}(t+1), & if \ f(X^{Pr}(t+1)) < f(Lb^{Pr}(t)) \end{cases} \tag{5}$$

Assuming that the function f is to be maximized. In the swarm of particles, the global best position found so far can be obtained using (6).

$$Gb^{Pr}(t+1) = \begin{cases} arg \ \max_{Lb^{Pr}(t+1)} f(Lb^{Pr}(t+1), & if \ \max f\left(Lb^{Pr}(t+1) \geq f(Gb^{Pr}(t)\right) \\ Gb^{Pr}(t), & if \ \max f\left(Lb^{Pr}(t+1) < f(Gb^{Pr}(t)\right) \end{cases} \tag{6}$$

We can use this binary form of the PSO to solve the binary combinatorial optimization problems. Our coverage problem for improve the lifetime in wireless sensor network can be formulated as a binary optimization problem and we can use the Binary PSO (BPSO) to solve it.

## 3.3  Network Model

Static sensor devices are uniformly deployed with high density in the target sensing field to ensure the coverage in the target sensing field. Homogeneous sensor devices are considered in this paper. It is assumed that every sensor device knows its position in the sensing field using either a GPS or a position discovery software. The wireless sensor node will use the binary disc sensing model by which each sensor node will has an appropriate sensing range is reserved within a circular disk called radius Rs. Each location within the area of interest is said to be covered if it lies within at least one sensor's sensing range. It is supposed that the range of communication $R_c \geq 2R_s$, where Zhang and Hou proved in them article [35] that if the range of communication is greater than or equal to the range of sensing. A complete coverage of the target sensing field leads to ensure the connectivity among the active sensors' nodes.

## 4  Coverage Problem Formulation

Given the interested sensing field A, the wireless sensor nodes set $S = \{s_1, \ldots, s_N\}$ that are deployed uniformly in this area such that they are ensure a full coverage for A. The area of the target sensing field A is partitioned into subfields $A = \{A^1, \ldots, A_R^N\}$, where $N_R$ is the number of subfields in A. We suppose that each sensor node $s_i$ know its position and its subfield. The subset $\delta k = \{s_1, \ldots, s_{N^k}\}$ represents the sensor nodes inside the subfield k in A. Each sensor node si has the same initial energy $E_i$ in the first time and the current residual energy $RE_i$ equal to $E_i$ in the first time for each $s_i$ in A. We define $\rho^k = \{P_1, \ldots, P_{N^k}\}$, where $N^k$ is the set of the points of interest that represents all the sensor nodes in the monitored subfield $A^k$ to satisfy the coverage requirement in each subfield. The node positions represent the points of interest in each subfield $A^k$. We can formulate our optimization problem as energy cost minimization by minimize the number of active sensor nodes and maximizing the coverage rate at the same time in each subfield $A^k$. The decision variable $D_j$ is formulated as in (7):

$$D_j = \begin{cases} 1, & \text{if } s_j \text{ is active during round } t(t = 1, \ldots, T) \\ 0, & \text{Otherwise.} \end{cases} \tag{7}$$

We can define the points of interest coverage indicator as follow

$$\Gamma_{jp} = \begin{cases} 1, & \text{if the point of interest } P_p \text{ is covered by active sensor node } s_j \\ 0, & \text{Otherwise.} \end{cases} \tag{8}$$

The coverage indicator $\Gamma_{jp}$ can be used to determine the redundancy of coverage of each point of interest in $PSET^k$ of $A^k$ that must be minimized from the produced list of active nodes $D_j$. Our distributed coverage optimization problem can be formulated as follow

$$Minimize \sum_{p=1}^{N^k} \sum_{j=1}^{N^k} \Gamma_{jp} * D_j \tag{9}$$

**Subject to**

$$\sum_{p=1}^{N^k} \sum_{j=1}^{N^k} \Gamma_{jp} * D_j = N^k \tag{10}$$

$$D_j \in \{0, 1\}, \forall s_j \in A^k n \tag{11}$$

In Eq. (9), the main objective is to minimize the number of active sensor nodes in the produced final solution vector D which leads to improve the lifetime of wireless sensor network. This goal is constrained by means of covering each point of interest in $SSET^k$ by at least one sensor active node. This refer to that our coverage optimization problem is a multi-objective optimization problem and we can use the BPSO to solve it. The basic coverage model inspired from [8] but we use it with some modification to be applied later by BPSO.

## 5   Distributed BPSO Algorithm for Energy-Efficient Area Coverage

Our Distributed BPSO Algorithm will be distributed and implemented in each subfield Ak simultaneously so as to solve the coverage optimization problem in centralized way by the cluster head in each subfield. The network lifetime for each subfield is divided into two stages: the discovery stage and the working stage. Figure 1 shows the main phases of our proposed EPSOLCOP protocol. This protocol can be implemented in each subfield as follow.

### 5.1   Discovery Phase

In the wireless sensor network initialization, each sensor node sends its position and remaining energy to all wireless sensor nodes in its subfield $A^k$ by using INFO packet and listen to the packets sent from other nodes in $A^k$. After that, each node will have information about all the sensor nodes in the region. The neighbor's discovery phase is performed only once during the network initialization at each region.

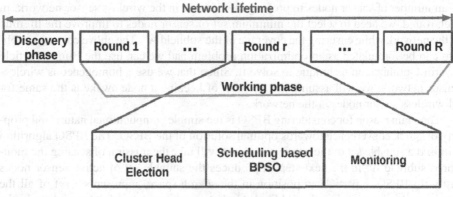

**Fig. 1.** EPSOLCOP protocol.

### 5.2   Working Phase

In this phase, the multi-objective distributed coverage protocol will start working by dividing its work into R rounds where each round will include four steps that can be explained as follow:

### 5.2.1   Cluster Head Selection Step

Cluster Head Selection Step: This step includes choosing the cluster head of the subfield which will be responsible of executing the Binary Particle Swarm Optimization (BPSO) algorithm in the next step to choose the list of active sensor nodes that contribute in covering the subfield $A^k$. The sensors in the same subfield capable of communicating with each other's using a multi-hop communication protocol. After the discovery phase each node will have the information of all other sensor nodes in the subfield and by using this information, each node will choose the cluster head based on the remaining energy $E_j$ of the sensors inside the subfield. If we have more than one node has the same $E_j$, this leads to choose the cluster head based on the largest index among them.

Each subfield Ak in the area of interest A will select its cluster head independently and at each round and this cluster head will execute the BPSO to gives the best representative sensor nodes that cover the subfield $A^k$ with minimum number of nodes. After that, the cluster head will send Active-Sleep packet to each sensor $s_j$ in the subfield Ak based on the BPSO algorithm decision vector X. The cluster head will stay active or go to Sleep based on the BPSO decision vector until the beginning of the next round. The next round starts with remaining energy exchange among the sensor nodes inside each subfield in A and then cluster head election.

### 5.2.2    Scheduling Based BPSO Step

The main goal in this step after choosing the cluster head is to produce the best active nodes set that will take the responsibility of covering the whole subfield $A^k$ with minimum number of sensor nodes to prolong the lifetime in the wireless sensor network. In each round, we need to select the minimum set of sensor nodes to improve the lifetime of the network while ensuring the coverage of the subfield $A^k$. The above coverage problem can be formulated as an optimization problem and we can use the Binary Particle Swarm Optimization technique to solve it. Since that we use a homogeneous wireless sensor network, we will assume that the cost of keeping a node awake is the same for all wireless sensor nodes in the network.

The main reason for considering BPSO is the simple computational nature and property of quick convergence towards optimal solution of the BPSO. The BPSO algorithm will find a suitable set of the active nodes that will take the mission of sensing the monitored subfield $A^k$ in the next step. It produces the suitable set of active sensor nodes fastly. In BPSO, a particle (a position in the search space) represents a set of all the sensor nodes for covering the subfield $A^k$ that represents a candidate solution in the search space of the problem. Each particle Pr in the swarm will includes position $X_j^{Pr}$, velocity $V_j^{Pr}$, local best $Lb_j^{Pr}$, global best $Gb_j$, where $1 \leq Pr \leq P_{size}$ and $1 \leq j \leq N^k$. The particle Pr's position can be represented as the vector $\{X_1,..., X_N^k\}$. Each value in the vector X maybe 1 if $s_j$ is active in $A^k$ or 0 for sleep state. We can initialize the position vector X using Eq. (12)

$$X_j^{Pr} = \begin{cases} 1, & \text{if rnd} > 0.5 \\ 0, & \text{Otherwise.} \end{cases} \tag{12}$$

Where rnd is a random number $\in [0, 1]$. Every velocity vector V of every particle Pr is initiated randomly with a restricted value using Eq. (13).

$$V_j^{Pr} = \text{Vmin} + (\text{Vmax} - \text{Vmin}) * \text{rnd} \tag{13}$$

Where $V_{min}$ and $V_{max}$ represents the minimum and the maximum velocity values respectively. We will update the velocity vector using (1) and after each update the resulted values for the velocity vector will be restricted to $[V_{min}, V_{max}]$ using Eq. (14)

$$f(x) = \begin{cases} Vmax, & if \ V_j^{Pr} > Vmax \\ Vmin, & if \ V_j^{Pr} < Vmin \\ V_j^{Pr}, & Otherwise. \end{cases} \tag{14}$$

Our BPSO used Eqs. (15), (16), and (17) to update $\omega$, c1, and c2 respectively.

$$\omega = (start - end) * ((MAXG - t)/MAXG) + end \tag{15}$$

$$c1 = (end - start) * ((MAXG - t)/MAXG) + start \tag{16}$$

$$c2 = (start - end) * ((MAXG - t)/MAXG) + end \tag{17}$$

In addition, we initialized Gfit and $Lfit^{Pr}$ to the worst possible values. Each particle can be evaluated by using the fitness function $F^{Pr}$ as in Eq. (18).

$$F^{Pr} = \sum_{p=1}^{N^k} \sum_{j=1}^{N^k} \Gamma_{jp} * D_j \tag{18}$$

One of the main problems that the BPSO usually suffers from it is the premature convergence. Sometimes, the BPSO converge to the local optimal solution instead of global optimal. In order to overcome these shortcomings and get better solution, we used the modified version of the position update equation in [4] instead of the sigmoid function (4) to update the position vector as in relation (19)

$$X_j^{Pr} = \begin{cases} 1, & if \ rnd < \frac{X_j^{Pr} + V_j^{Pr} + V_{max}}{1 + 2*V_{max}} \\ 0, & Otherwise. \end{cases} \tag{19}$$

The relation (19) allows for better exploration to the search space that leads the BPSO to gives optimal or near optimal solution. The proposed BPSO algorithm to solve our coverage problem is presented in Algorithm 1.

The time complexity for our BPSO algorithm is $O(MAXG * P_{size} * N^k * N^k)$ in worst case analysis. The outer loop (line 9) need as upper bound equal to $O(MAXG)$ and it can be decreased based on the stopping criteria. The inner loop that related to update local and global best (lines 10–14) needs $\theta(MAXG * P_{size} * N^k)$ in the best and worst-case analysis. The last inner loop (lines 16–20) that Responsible for update the velocity and the position of particle will takes as best and worst cases equal to $\theta(P_{size} * N^k * N^k)$.

### 5.2.3 Monitoring Step

The BPSO algorithm will produce the best set of the active nodes that will take the mission of covering the subfield $A^k$ in monitoring step. After the convergence of the BPSO to the optimal (or near optimal) solution, the cluster head will send Active-Sleep Packet to all sensor nodes in Ak to tell them who will stay active or sleep a round of time fixed for the monitoring phase until the starting time of the next round.

---

**Algorithm 1** BPSO for Coverage Problem

---

**Require:** all the parameters that related to both coverage protocol and the BPSO algorithm
**Ensure:** X (solution vector that represents the states of sensor nodes in the next step)
1: Compute the parameter $\Gamma_{jp}$ using equation (8);
2: **for** $Pr \leftarrow 1$ **to** $Psize$ **do**
3:     Initialize $X^{Pr}$ randomly by 0 or 1 using equation (12);
4:     Initialize $V^{Pr}$ using equation (13) ;
5:     Compute the fitness value $f(X^{Pr})$ using equation (18) ;
6:     Initialize $Lb^{Pr} \leftarrow X^{Pr}$;
7:     Initialize $f(Lb^{Pr}) \leftarrow -\infty$ ;
8: **end**
9: Initialize $f(Gb) \leftarrow -\infty$ ;
10: **While** (t $\leq$ MAXG) and (The termination condition is not met) **do**
11:     **for** $Pr \leftarrow 1$ **to** $Psize$ **do**
12:         **if** $f(X^{Pr}) > f(Lb^{Pr})$ **then**
13:             $Lb^{Pr} \leftarrow X^{Pr}$;
14:         **end**
15:         **if** $f(Lb^{Pr}) > f(Gb)$ **then**
16:             $Gb \leftarrow Lb^{Pr}$;
17:         **end**
18:     **end**
19:     Update $\omega$, $c_1$, and $c_2$ using Equations (15), (16) and (17) respectively;
20:     **for** $Pr \leftarrow 1$ **to** $Psize$ **do**
21:         Update the Velocity $V^{Pr}$ using Equation(1);
22:         Restrict the Velocity $V^{Pr}$ using Equation(14);
23:         Update the Position $X^{Pr}$ using Equation(19);
24:         Compute the quality of solution $X^{Pr}$ using Equation (18);
25:     **end**
26:     $t \leftarrow t + 1$;
27: **end**
28: **return** $X$;

---

## 6 Performance Evaluation and Analysis

This section evaluates the efficiency of proposed EPSOLCOP protocol by conducting the results using the network simulator OMNeT++ [29]. The parameters of the simulation are reviewed in Table 1.

The conducted results are the average of 50 simulation runs for various topologies of the network. The network simulation is achieved on five different WSN sizes ranging from 100 to 300 nodes. The sensor nodes in these networks are deployed uniformly in the target sensing field which is $(50 \times 25)$ m$^2$ to ensure the coverage over the whole WSN with the given sensing range. Our protocol is performed in the sensing field of the WSN which is divided into 16 regular subfields. Our protocol employed this network subdivision way and the energy consumption model which are described in [15]. Of course, this number of subfields should be adapted according to the size of the sensing field area of interest and the number of sensors.

**Table 1.** Parameters values for network initializing.

| Parameter | Value |
|-----------|-------|
| Sensing field | $(50 \times 25)$ m$^2$ |
| Nodes number | 100, 150, 200, 250, and 300 nodes |
| Initial energy | 500–700 J |
| Sensing period | One hour |
| $E_{thr}$ | 36 J |
| $R_s$ | 5 m |
| $R_c$ | 10 m |
| MAXG | 1000 |
| Psize | 30 |

Each sensor device is initialized with a random energy value within the range [500–700]. The minimum energy required by the sensor node during one round is $E_{th} = 36$ J. Every node in the WSN will not continue to the next round if it has residual energy less than $E_{th}$.

Our proposed protocol is evaluated using the same performance metrics that employed by [15] such as Coverage Ratio (CR), Active Sensors Ratio (ASR), Network Lifetime, and Energy Consumption. The proposed protocol is executed on portable computer DELL Intel Core i3 2370 M (1.8 GHz) processor (2 cores). The MIPS (Million Instructions Per Second) rate is equal to 35330. In order to be compatible with using a sensor device that has Atmels AVR ATmega103L microcontroller (6 MHz) with MIPS rate equal to 6, the calculated execution time on the portable computer is multiplied by 2944.2 ($\frac{35330}{2} * \frac{1}{6}$).

Besides EPSOLCOP, two existing related works will be evaluated for comparison purposes. The first algorithm, named DESK, is a fully distributed coverage algorithm suggested by [30]. The second algorithm, named GAF [32], which di-vides the monitoring area into fixed squares. Then, during the decision phase, each square, one sensor device is selected to stay active throughout the monitoring phase. The proposed EPSOL-COP protocol attempts to ensure the coverage of the whole the points of interest in each subfield inside the target sensing field. EPSOLCOP protocol uses BPSO to optimize both coverage and network lifetime in WSN.

## 6.1  Coverage Ratio

In this section, the average coverage ratio produced by our proposed EPSOL-COP protocol and for 200 deployed nodes is studied (see Fig. 2). The result shows that DESK and GAF give a slightly better ratio of coverage with respectively 99.99% and 99.91% compared to the 98.09% produced by EPSOLCOP protocol for the first rounds. EPSOL-COP protocol turns off larger redundant sensor nodes at the beginning of the network lifetime that slightly reduces the coverage ratio whilst DESK and GAF activates a larger number of sensor nodes during each round because they did not optimize the coverage and the lifetime in the WSN. After that, when the number of rounds is beyond 44, it is obviously that EPSOLCOP gives a better coverage ratio than other two methods and for longer rounds due to applying the optimization that saved the energy of sensor nodes and improve the lifetime of WSN. The energy saved by EPSOLCOP in the early rounds permits to this substantial increase in the performance of the coverage.

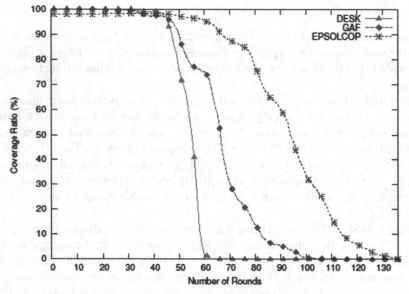

**Fig. 2.**  Coverage ratio for 200 deployed nodes.

## 6.2  Active Sensors Ratio

In each round, it is important to activate as less number as possible of sensor nodes so as to conserve the energy and extend the WSN lifetime. The average active nodes ratio for 200 deployed nodes are shown in Fig. 3. In the first fourteen rounds, DESK and GAF activate 30.36% and 34.96% sensor nodes whilst our proposed protocol activates 20.04% of sensor nodes during the same time interval. When the number of rounds increases, our proposed protocol activates a larger number of sensor devices compared to DESK, and GAF due to the decreased number of dead nodes and the capability of the protocol to provide a higher ration of coverage as shown in Fig. 2.

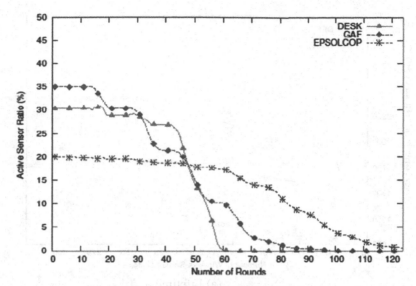

**Fig. 3.** Active sensors ratio for 200 deployed nodes.

## 6.3 Energy Consumption

Figure 4(a) and (b) show the energy consumption for different network densities and for $Lifetime_{95}$ and $Lifetime_{50}$. We study in this section the impact of the consumed energy by the sensor nodes for various WSN sizes. The simulation results show that our proposed EPSOLCOP consumes less energy compared with the other two methods due to activating a lower number of sensor nodes during each round using the scheduling-based optimization in the decision phase. Figure 4(a) and (b) illustrate the superiority of our proposed EPSOLCOP protocol from the energy saving point of view.

## 6.4 Network Lifetime

In Fig. 5(a) and (b), $Lifetime_{95}$ and $Lifetime_{50}$ are shown for different network densities. The simulation results show that EPSOLCOP protocol improves the network lifetime for both $Lifetime_{95}$ and $Lifetime_{50}$ in comparison with other two approaches due to activating a lower number of sensor nodes and consuming a less energy in each round. As shown in these figures, the lifetime increases when the size of the network increases.

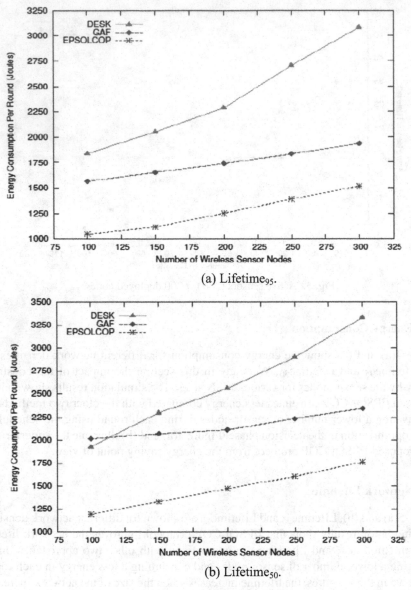

(a) Lifetime$_{95}$.

(b) Lifetime$_{50}$.

**Fig. 4.** Energy consumption per period.

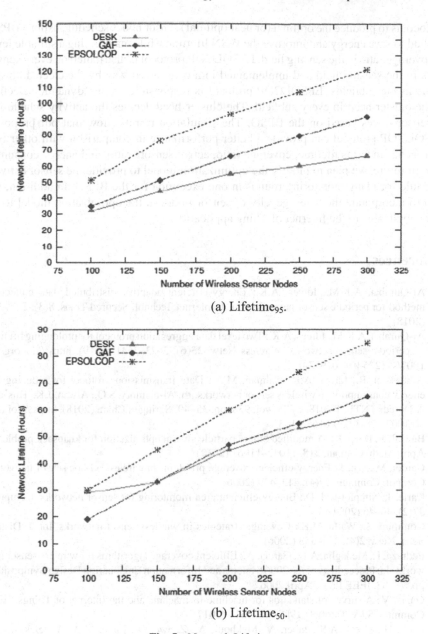

(a) Lifetime$_{95}$.

(b) Lifetime$_{50}$.

**Fig. 5.** Network Lifetime.

## 7   Conclusions and Future Works

This article studied the problem of lifetime coverage optimization in WSNs. We propose a new protocol named Energy-efficient Particle Swarm Optimization for Lifetime Coverage Prolongation (EPSOLCOP) in WSNs. The main goal of our proposed EPSOLCOP protocol is to produce the optimal (or near optimal) sensor nodes schedule using a BPSO method to save energy and improve the WSN lifetime while maintaining a suitable level of coverage ratio in the sensing field. EPSOLCOP protocol is distributed on every sensor node in the sensing field and implemented in a centralized way by the elected cluster heads in the subfields. EPSOLCOP protocol is responsible for the dynamic selection of the cluster head in every subfield. The cluster head decides the activity scheduling of sensor nodes based on the BPSO. The simulation results show that our proposed EPSOLCOP protocol can provide a better performance in comparison with other two approaches in terms lifetime, coverage ratio, active sensors ratio, and energy consumption. In future, we plan to modify the optimization model to provide the sensor activity schedule for many monitoring rounds in one execution for the BPSO. In addition, we plan to incorporate the heterogeneity of sensor nodes in the optimization model to be more applicable to the Internet of Thing applications.

# References

1. Al-Qurabat, A.K.M., Idrees, A.K.: Energy-efficient adaptive distributed data collection method for periodic sensor networks. Int. J. Internet Technol. Secured Trans. **8**(3), 297–335 (2018)
2. Al-Qurabat, A.K.M., Idrees, A.K.: Two level data aggregation protocol for prolonging lifetime of periodic sensor networks. Wireless Netw. **25**(6), 3623–3641 (2019). https://doi.org/10.1007/s11276-019-01957-0
3. Alhussaini, R., Idrees, A.K., Salman, M.A.: Data transmission protocol for reducing the energy consumption in wireless sensor networks. In: Al-mamory, S.O., Alwan, J.K., Hussein, A.D. (eds.) NTICT 2018. CCIS, vol. 938, pp. 35–49. Springer, Cham (2018). https://doi.org/10.1007/978-3-030-01653-1_3
4. Bansal, J., Deep, K.: A modified binary particle swarm optimization for knapsack problems. Appl. Math. Comput. **218**, 11042–11061 (2012)
5. Cardei, M., Wu, J.: Energy-efficient coverage problems in wireless ad-hoc sensor networks. Comput. Commun. **29**(4), 413–420 (2006)
6. Carle, J., Simplot-Ryl, D.: Energy-efficient area monitoring for sensor networks. Computer **37**(2), 40–46 (2004)
7. Commuri, S., Watfa, M.K.: Coverage strategies in wireless sensor networks. Int. J. Distrib. Sens. Netw. **2**(4), 333–353 (2006)
8. Pedraza, F., Medaglia, A.L., Garcia, A.: Efficient coverage algorithms for wireless sensor networks. In: Proceedings of the 2006 Systems and Information Engineering Design Symposium, pp. 78–83. IEEE Press, April 2006
9. Gazis, V.: A survey of standards for machine-to-machine and the Internet of Things. IEEE Commun. Surv. Tutorials **19**(1), 482–511 (2017)
10. Harb, H., Idrees, A.K., Jaber, A., Makhoul, A., Zahwe, O., Taam, M.A.: Wireless sensor networks: a big data source in internet of things. Int. J. Sens. Wireless Commun. Control **7**(2), 93–109 (2017)
11. Huang, C.F., Tseng, Y.C.: The coverage problem in a wireless sensor network. Mob. Netw. Appl. **10**(4), 519–528 (2005). https://doi.org/10.1007/s11036-005-1564-y
12. Idrees, A.K., Al-Qurabat, A.K.M.: Distributed adaptive data collection protocol for improving lifetime in periodic sensor networks. IAENG Int. J. Comput. Sci. **44**(3), 345–357 (2017)
13. Idrees, A.K., Al-Yaseen, W.L.: Distributed genetic algorithm for lifetime coverage optimization in wireless sensor networks. Int. J. Adv. Intell. Paradigm (in press)

14. Idrees, A.K., Al-Yaseen, W.L., Taam, M.A., Zahwe, O.: Distributed data aggregation based modified k-means technique for energy conservation in periodic wireless sensor networks. In: 2018 IEEE Middle East and North Africa Communications Conference (MENACOMM), pp. 1–6. IEEE (2018)
15. Idrees, A.K., Deschinkel, K., Salomon, M., Couturier, R.: Distributed lifetime coverage optimization protocol in wireless sensor networks. J. Supercomput. **71**(12), 4578–4593 (2015). https://doi.org/10.1007/s11227-015-1558-x
16. Idrees, A.K., Deschinkel, K., Salomon, M., Couturier, R.: Multiround distributed lifetime coverage optimization protocol in wireless sensor networks. J. Supercomput. **74**(5), 1949–1972 (2018)
17. Idrees, A.K., Deschinkel, K., Salomon, M., Couturier, R.: Perimeter-based coverage optimization to improve lifetime in wireless sensor networks. Eng. Optim. **48**(11), 1951–1972 (2016)
18. Kennedy, J., Eberhart, R.: A discrete binary version of the particle swarm algorithm. In: Proceedings of the International Conference on Systems, Man, and Cybernetics. vol. 5, pp. 4104–4108, October 1997
19. Kennedy, J., Eberhart, R.C.: Particle swarm optimization. In: Proceedings of IEEE International Conference on Neural Networks, vol. 4, pp. 1942–1948, December 1995
20. Kumar, S., Lai, T.H., Arora, A.: Barrier coverage with wireless sensors. In: MobiCom 2005 Proceedings of the 11th Annual International Conference on Mobile Computing and Networking, pp. 284–298. ACM Press, August 2005
21. Li, J., Cheng, S., Gao, H., Cai, Z.: Approximate physical world reconstruction algorithms in sensor networks. IEEE Trans. Parallel Distrib. Syst. **25**(12), 3099–3110 (2014)
22. Mostafaei, H., Shojafar, M.: A new meta-heuristic algorithm for maximizing lifetime of wireless sensor networks. Wireless Pers. Commun. **82**(2), 723–742 (2015). https://doi.org/10.1007/s11277-014-2249-2
23. Mulligan, R., Ammari, H.M.: Coverage in wireless sensor networks: a survey. Netw. Protoc. Algorithms **2**(2), 27–53 (2010)
24. Nan, G., Shi, G., Mao, Z., Li, M.: CDSWS: coverage-guaranteed distributed sleep/wake scheduling for wireless sensor networks. EURASIP J. Wireless Commun. Netw. **2012**(1), 44 (2012). https://doi.org/10.1186/1687-1499-2012-44
25. Özdemir, S., Bara'a, A.A., Khalil, Ö.A.: Multi-objective evolutionary algorithm based on decomposition for energy efficient coverage in wireless sensor networks. Wireless Pers. Commun. **71**(1), 195–215 (2013). https://doi.org/10.1007/s11277-012-0811-3
26. Rebai, M., Le Berre, M., Hnaien, F., Snoussi, H., Khoukhi, L.: A branch and bound algorithm for the critical grid coverage problem in wireless sensor networks. Int. J. Distrib. Sens. Netw. **10**, 769658 (2014)
27. Shi, Y., Eberhart, R.C.: A modified particle swarm optimizer. In: Proceedings of the IEEE Congress on Evolutionary Computation, pp. 69–173, May 1998
28. Thai, M.T., Wang, F., Du, D.H., Jia, X.: Coverage problems in wireless sensor networks: designs and analysis. Int. J. Sens. Netw. **3**(3), 191–200 (2008)
29. Varga, A.: Omnet++ discrete event simulation system (2003). http://www.omnetpp.org
30. Vu, C., Gao, S., Deshmukh, W., Li, Y.: Distributed energy-efficient scheduling approach for k-coverage in wireless sensor networks. In: MILCOM, pp. 1–7 (2006). http://doi.ieeecomputersociety.org/10.1109/MILCOM.2006.302146
31. Wang, B.: Coverage problems in sensor networks: A survey. ACM Comput. Surv. (CSUR) **43**(4), 32 (2011)
32. Xu, Y., Heidemann, J., Estrin, D.: Geography-informed energy conservation for ad hoc routing. In: Proceedings of the 7th Annual International Conference on Mobile Computing and Networking, pp. 70–84. ACM (2001)

33. Yu, J., Ren, S., Wan, S., Yu, D., Wang, G.: A stochastic k-coverage scheduling algorithm in wireless sensor networks. Int. J. Distrib. Sens. Netw. **8**(11), 746501 (2012)
34. Yu, J., Wan, S., Cheng, X., Yu, D.: Coverage contribution area-based k-coverage for wireless sensor networks. IEEE Trans. Veh. Technol. **66**, 8510–8523 (2017)
35. Zhang, H., Hou, J.C.: Maintaining sensing coverage and connectivity in large sensor networks. Ad Hoc Sens. Wireless Netw. **1**, 89–124 (2005)

# Iris Recognition Using Localized Zernike Features with Partial Iris Pattern

Sinan A. Naji[1]([⊠])([iD]), Robert Tornai[2], Jasim H. Lafta[2], and Hussein L. Hussein[3]

[1] Department of Postgraduate Studies, University of Information Technology and Communications, Baghdad, Iraq
dr.sinannaji@uoitc.edu.iq
[2] Doctoral School of Informatics, Faculty of Informatics, University of Debrecen, Debrecen, Hungary
tornai.robert@inf.unideb.hu, jasimhussein67@gmail.com
[3] Department of Computer Science, Ibn Al-Haitham College of Education for Pure Sciences, University of Baghdad, Baghdad, Iraq
hussein.l.h@ihcoedu.uobaghdad.edu.iq

**Abstract.** Iris recognition is an attractive field of research for the purpose of identifying people based on unique patterns extracted from their iris. Accurate iris recognition requires high quality of iris images with the most discriminating features and minimum variation level. Analyzing and encoding of an iris image captured by a real (i.e., non-ideal) imaging system is a crucial issue. This paper presents a new iris pattern for human identification; i.e., partial iris pattern, instead of the whole iris. Then, a feature-set consisting of Gabor and Zernike moments is used for feature extraction and matching. The features are calculated to the ninth order. Two datasets are used for testing and evaluating the proposed technique. The experiential results show that the proposed technique is reliable, promising, and can be applied in real world imaging systems.

**Keywords:** Iris recognition · Zernike moments · Partial iris pattern · Feature extraction

## 1 Introduction

Person identification and authentication are daily procedures for several kinds of activities such as access gates to gain entry into restricted areas, ATM machines, police, immigration, equipment usage, etc. In traditional systems, the procedure is based on something that one knows (such as user-name and passwords), or something that one carries (such as magnetic-cards, keys, or chip-cards). As technologies and services are becoming available everywhere, these methods are no longer secure. The passwords can be attacked. Cards may be stolen. To achieve secure human identification as well as reduce the fraudulent claims of individuals, we should use something that actually discriminates each person. Biometrics technology offers an efficient way for human identification based on measurable physiological and/or behavioral characteristics of

A. M. Al-Bakry et al. (Eds.): NTICT 2020, CCIS 1183, pp. 219–232, 2020.
https://doi.org/10.1007/978-3-030-55340-1_16

individuals. These characteristics may include fingerprints, human face, hand geometry, iris, signature dynamics, voice, etc. It has been proven that these characteristics are unique for each individual and can be used efficiently for human identity verification [1]. Currently, the existing state-of-the-art systems provide reliable and accurate automatic identification, and this field is still a hot topic for researchers to provide high level of automation, accuracy, and fast processing.

Unfortunately, these methods are quite intrusive and have not gained acceptance by ordinary people for two main reasons: 1) Individuals have to give samples of their characteristics every time; and this is really not preferred. 2) Individuals are required to use equipment also being used by others; and this is not acceptable by many.

Although automatic face recognition is the most acceptable application of biometric identification systems which have gained significant attention due to the fact that it is non-intrusive [1], it has been proven that even for "mug shot" images, the existing optimal algorithms can have inevitable error rates due to the key issue of inter-class and intra-class variations [2]. For example, the challenges arise from the fact that a human face is a three-dimensional object projected into 2D image can cause significant differences in face appearance due to pose, occlusion, scaling, rotation, illumination or camera characteristics. For example, many studies have stated that variability among images of the same face due to illumination are almost larger than image differences of different faces captured under fixed illumination [3, 4].

Automatic iris biometric is yet another alternative for non-intrusive identity verification. Recently, many studies have shown that iris recognition is becoming a more reliable visual-based recognition with high confidence level especially when dealing with large databases due to the following reasons [2, 5–8]: (1) The iris of the human eye provides an extremely rich-data structure that is unique to each individual (i.e., it contains distinctive features such as furrows, vasculature, rings, crypts, freckles, etc.); (2) the iris is more stable over aging, and well protected over time; (3) iris images are less sensitive to unconstrained imaging (i.e., the iris pattern can be encoded from distances of up to one meter that can be done without individual's cooperation or even prior knowledge using a hidden-camera); (4) ease of iris localizing with fast preprocessing operations such as scaling, rotation, translation, and alignment (i.e., aligning all iris images so that the approximate positions of the iris features are the same in all images) [5, 9].

Figure 1 shows that the distinctiveness of the iris texture is clear even for casual users. Figure 1(a) shows three iris images of the same individual. Figure 1(b) shows three iris images of different individuals. As shown in these figures, the iris patterns of different irises are markedly distinct.

Generally, the term *automatic iris-recognition* can be defined as follows: Given an arbitrary person, the goal of iris recognition is to use the iris of the human eye for identity verification by analyzing and encoding the texture of the iris. However, as any human identification system, iris biometric systems are also exposed to fraud attacks such as paper printed iris, cosmetic contact lens, replayed videos, etc.

An important step to improve the accuracy of the system is to reduce the amount of variation between iris images of the dataset. This would give the most compact space of iris images. To the best of our knowledge, all previous works have used whole iris pattern to analyze and encode the iris. One of the challenging problems is the high variability

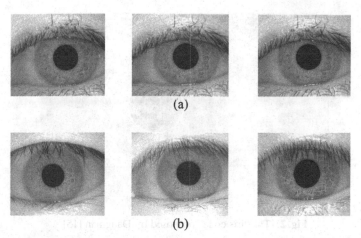

(a)

(b)

**Fig. 1.** The distinctiveness of the iris texture is clear even for casual users; (a) Iris images of the same individual; (b) Iris images of different individuals.

in images captured in unconstrained environment that makes iris images non-ideal due to many reasons such as presence or absence of structural components (i.e., eyelashes, eyelid, hair, etc.). It is clear that reducing image variability will be helpful.

This study presents a novel hybrid system to analyze and encode iris images independent of scale, position, rotation and occlusion. The main idea is based on using partial iris pattern instead of the whole iris. With this solution, the system can cope with the problem of non-ideal iris images that makes texture analyzing and encoding easier, faster, and with low misrecognition rate. Furthermore, Zernike moments are used along with 2D-Gabor features to enhance the accuracy.

The remaining sections of this study are organized as follows: Sect. 2 presents the background. Section 3 describes the proposed system. Section 4 shows the experimental results; and Sect. 5 is the conclusion.

## 2  Related Work

The earliest attempt for automatic iris-recognition system as a basis for human identity verification goes back to Flom and Safir in 1987 [10]. Although it is not clear that this attempt implemented and tested a real working system, it nevertheless proposed the general concepts of iris recognition. In 1991, Johnson R. G. [11] presented a study that actually realized a system for automated iris recognition. Then, an interesting work was presented by John Daugman from Cambridge University in 1992 and 1993 [12, 13]. Daugman proposed an efficient method for transforming the extracted features of the iris texture into numerical code using 2D Gabor wavelets (Fig. 2). Then, Hamming distance was used for matching. Further research has been carried out to improve the performance [14, 15]. As reported by authors, the performance is highly reliable and makes it an ideal method for visual human identification. To the best of our knowledge, many commercially available systems are based on the Daugman'smethod.

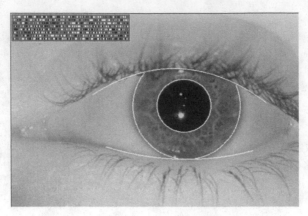

**Fig. 2.** The "iris code" proposed by Daugman [15].

Wildes [7] used Gaussian filter (in both the vertical and horizontal directions) at different image scales for iris encoding and proposed the normalized correlation as a similarity metric for matching. Boles and Boashash[16] proposed an algorithm based on zero-crossing of wavelet transform at different scales. The output pattern is compared using multi-dissimilarity functions. Chu and Chin [17] proposed Sobel operator and wavelet transform for feature extraction. The system combines Probabilistic Neural Network (PNN) and Particle Swarm Optimization (PSO) for iris classification. Hosny and Hafez [18] proposed the use of Zernike moments for iris recognition. Later, Kuar et al. [5] and Pirasteh et al. [19] also suggested Zernike moments in their work. The main advantage of Zernike moments is that they do not contain redundant information [20]. Chen and sun [21] proposed a new descriptor, called the Zernike moment phase-based descriptor, where the phase information of a signal is more informative than the magnitude information. Rai et al. [22] proposed Haar wavelet decomposition and log-Gabor filters for feature extraction; and Support Vector Machine (SVM) along with Hamming distance for iris classification. Liu et al. [23] proposed an iris recognition model based on 2D-Gabor features and key points generated by Scale-Invariant Feature Transform (SIFT) algorithm. Then, a fusion rule is used to make the classification decision. Bhateja et al. [24] proposed an iris recognition model based on combining three classifiers to reduce the False Acceptance Rate (FAR). The proposed system uses sparse representation with k-nearest space model, sector based model, and cumulative concentration index model. The weighted outputs of the classifiers are used for final classification decision. Many other works on iris recognition have been presented by [5, 6, 25, 26].

## 3   The Proposed System

In practice, an iris-recognition system comprises the following four main steps: image capturing, pre-processing, extracting the features (i.e., feature encoding), and finally matching the extracted features with iris database. This implies that an iris image database should be collected and indexed in a systematic way. Nowadays, there are many iris databases that are publically available for researchers such as CASIA-iris database [27],

MMU iris database [28], WVU-Iris Database [29], etc. In most databases, the iris image does not contain only the iris but it captures the whole eye and part of the surrounding eye region. Therefore, before performing iris feature extraction, preprocessing tasks such as pupil and iris localization, scaling, rotation, translation, and alignment are required first. Figure 3 shows iris image examples used in this study from CASIA-iris database [27].

**Fig. 3.** Ideal iris images, from CASIA-iris database [27].

## 3.1  Iris and Pupil Localization

The iris-recognition system has to locate the region of interest (ROI) by localizing the pupil and iris on the eye-image. In an uncontrolled environment, image capturing cannot be expected to produce an image that encloses only the iris. Practically, we are dealing with an eye image that contains the entire eye and part of the surrounding regions. So, iris localization is a critical step that should be capable of dealing with variable occlusion, low intensities, edges, and so on. If certain parts of the iris are occluded (e.g., by eyelids or hair), the system must correctly locate the iris regions no matter the type of occlusion and returns the corresponding parameters modeled by the center coordinates $x_c$ and $y_c$, and the radius $r$. These parameters define the circle equation:

$$x_c^2 + y_c^2 - r^2 = 0 \tag{1}$$

The pupil should also be located accurately where the pupil is typically darker than the iris. However, recent advances on iris recognition have proposed different methods to deal with images captured in unconstrained environments. In this work, the implemented method was adopted from the work of Masek [30] which is based on using Circle Hough Transform (CHT) along with Canny edge detector. This method is robust to detect circular objects in an image and insensitive to noise. So, the system returns six coordinates: $x_c$ and $y_c$, and the radius $r$ of pupil along with $x_c$ and $y_c$, and the radius $r$ of iris. Theoretically, it is supposed that the center coordinates of the pupil and the iris are the same. In practice, the algorithm returns different center coordinates. In this work, the average of centers coordinates is calculated and used for both. Figure 4 shows an example of iris and pupil localization results. For the subsequent steps, only that portion of the iris between the inner and the outer boundaries must be analyzed. However, false iris localization can hardly be avoided since we are dealing with unlimited sort of variations that may lead to unpredictable results.

Figure 5 shows an example of false iris localization.

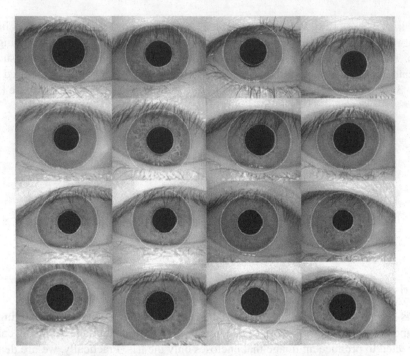

**Fig. 4.** The outline boundaries show results of the iris and pupil localization.

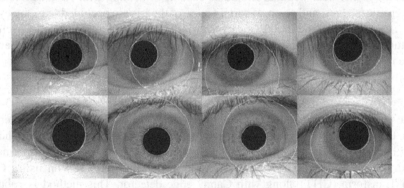

**Fig. 5.** False iris localization.

## 3.2 The Partial Iris Pattern

In practical experience, acquiring an iris image is a major issue for automated iris recognition. In fact, iris is a small object. Capturing iris images should be performed with high enough resolution, at-a-distance, under unconstrained conditions, and finally without disturbing people (because individuals are sensitive about their eyes). The challenges associated with iris image acquisition can be attributed to the following factors: scaling, rotations, translation, pose, illumination, occlusions (i.e., iris may be partially occluded by other objects such as eyelashes, eyelids, hair, eyeglasses). Furthermore, various factors may cause degradation during image capturing such as wrong focus, motion, camera characteristics, and random sensor errors.

Therefore, the quality of the training dataset is possibly one of the big obstacles for a successful application. A challenging issue affecting dataset quality is the high variability in iris images that can be attributed to the above-mentioned factors. To the best of our knowledge, there is inadequate evidence in the literature to address the limitations of iris images. Figure 6 shows some examples of non-ideal iris images where 30–55% of iris texture has been lost.

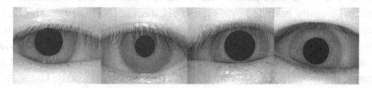

**Fig. 6.** Non-ideal iris images example (i.e., 30–55% of iris texture hasbeen lost).

The first step to improve the quality of iris images is to reduce the amount of iris variations between images. This would give the most compact space of iris images. To the best of our knowledge, most previous works have used whole iris patterns for feature extraction and encoding procedures. Therefore, the non-ideal iris images, such as those shown in Fig. 6 may degrade the system accuracy.

As a result of the general fact that the eye is oval-shaped while the real iris images are 2D rectangular images, one way to decrease iris image variability is by utilizing a binary mask. Generally, masking aims to exclude the areas with noise (i.e., eyelids, eyelashes, etc.). Practically, the value of 1's is used for useful areas (i.e., region of interest), and 0's elsewhere (i.e., non-iris). Excluding (or ignoring) non-iris areas from the training images will ensure that the system does not wrongly introduce any unwanted structures into the iris representation (reducing iris variability). The shape of the mask can be variable or fixed. Figure 7 shows a sample of fixed shape masking proposed by [30]. Figure 8 shows a sample of variable masking proposed by [31].

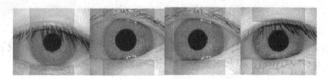

**Fig. 7.** Fixed shape masking used for excluding areas with noise, proposed by [30].

Although masking technique improves the quality of iris samples, this solution still implies pattern variability drawback. Based on this conclusion, a novel iris pattern has been suggested, i.e. partial iris pattern, instead of the whole iris pattern. To the best of our knowledge, this is the first attempt that employs partial iris pattern for human

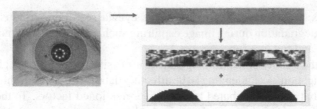

**Fig. 8.** Variable shape masking used for excluding areas with noise, proposed by [31].

identification. The basic idea behind that is as follows: the iris constitutes a circular shape that can be easily bounded in a square Iris Region IR (Fig. 9a). The IR region is divided into nine sub-regions (i.e., 3 × 3), where the pupil would be centered at IR(2, 2) as shown in Fig. 9(b). The goal of this step is to compute the standard deviation ($\sigma$) of each squared sub-region in order to figure out the ones that have the lowest variation for both cases: the same individual and then for different individuals. Consider that we have a set of iris images stacked at each other as shown in Fig. 10. There are $m$ pixels for any given coordinates $(i, j)$, one pixel at that location for each image.

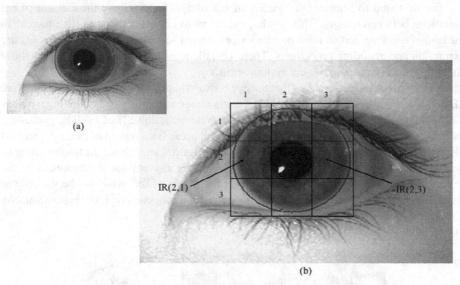

**Fig. 9.** The proposed partial iris pattern; (a) The iris pattern; (b) The iris region IR is divided into 3 × 3 sub-regions where the pupil is centered at IR(2,2) entry.

The average and standard deviation of each sub-region is calculated as follows:

**Step 1:** Obtain iris images $I_1, I_2,...,I_m$ (training dataset) where $m$ is the number of iris images. The iris images should be of the same size, with the pupil centered at IR(2, 2)
**Step 2:** Represent every image $I_i$ as a vector $\Gamma_i$
**Step 3:** Compute the average iris vector $\Psi$

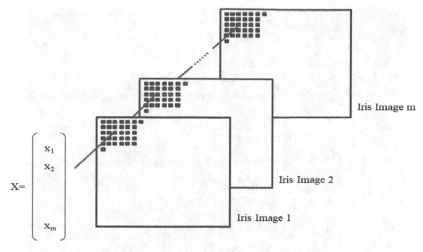

**Fig. 10.** Stack of iris images of the same size.

$$\Psi = \frac{1}{m} \sum_{i=1}^{m} \Gamma_i \tag{2}$$

**Step 4:** Compute the standard deviation vector ($\sigma$):

$$\sigma = \frac{1}{m} \sqrt{\sum_{i=1}^{m} (\Gamma_i - \Psi)(\Gamma_i - \Psi)^t} \tag{3}$$

**Step 5:** The standard deviation vector ($\sigma$) is presented as an iris image.

Experimentally, the calculated standard deviations revealed that the regions IR(2,1) and IR(2,3) show the lowest variations (see Fig. 9b). In pattern recognition problems, the regions that imply higher variations degrade the quality of the training data. Furthermore, it was found that about 50% of the cornersub-regionsare non-iris. So, we concluded that using partial iris pattern instead of whole iris pattern will improve the quality of the input data.

The partial iris pattern proposed in this study comprises IR(2,1) and IR(2,3) sub-regions. Figure 11 shows examples of the proposed partial iris pattern IR(2,1) region. Figure 12 shows examples of the proposed partial iris pattern IR(2,3). Compared to the binary masking technique, the partial iris pattern shows minimum area of noise or it is completely free of noise.

In practice, the iris encoding must be resistant to changes in the size of the iris, which depends on the distance to the eye and zoom. Thus, we must create an iris pattern that is invariant to its size in the eye image. So, after locating and extracting ROI, the size of partial iris pattern is fixed experimentally at 60 × 60 pixels to preserve sufficient

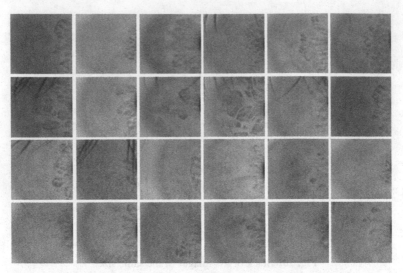

**Fig. 11.** The proposed partial iris pattern IR(2,1) region.

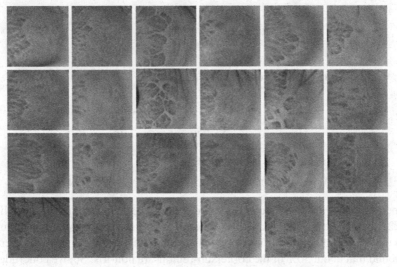

**Fig. 12.** The proposed partial iris pattern IR(2,3) region.

resolution to extract discriminated features and at the same time keep the dimensionality of the iris space manageably small.

In practice, using the partial iris pattern increases the True detection rate of the classifier. As the detection rate is increased, the number of false detections increased correspondingly. It is clear that suppressing false detections is an important issue. Therefore, more features are needed to improve the performance of the system. In this study, the Zernike moments are utilized to enhance the performance of the system.

### 3.3  Zernike Moments

After ROI is localized, processing of local features is needed for region description and encoding. Geometric transformations such as rotation, scaling, and translation are critical issues in most image analysis tasks [21]. A number of features and local descriptors have been proposed and investigated in the literature [32, 33]. From the point of view of pattern recognition, moment invariants are considered reliable features because of their promising performance if their values are insensitive to the presence of image noise [32].

However, the goal is to extract enough discriminating features which can deliver sufficient information for classification. For the feature extraction step, Gabor filters [14] along with Zernike moments (ZMs) [34] were used in our system. As the database comprises a large number of individuals, the proposed feature-set has the capability to generate enough invariant features. Zernike moments (ZMs) were originally introduced by Teague in the 1980s to reduce information redundancy existing in the standard geometric moments used for image classification. The ZMs are excellent region-based moments which are invariant to geometric transformation and can be extracted to an arbitrary order. Although the higher order moments are sensitive to noise, the ZMs attracted the attention of researchers. Nowadays, Zernike moments have been applied in many image analysis applications.

The ZMs basic function with order $n$ and repetition $m$ can be defined over a unit circle as follows [21, 30, 35]:

$$V_{nm}(x, y) = V_{nm}(\rho, \theta) = R_{nm}(\rho)e^{jm\theta} \tag{4}$$

$$R_{nm}(\rho) = \sum_{s=0}^{\frac{(n-|m|)}{2}} (-1)^s \frac{(n-s)!}{s!\left(\frac{(n+|m|)}{2} - s\right)!\left(\frac{(n-|m|)}{2} - s\right)!} \rho^{n-2s} \tag{5}$$

where $m$ and $n$ are integers such that $n$ is a non-negative, $(n-|m|)$ is even, $|m| \leq n$, $\rho = \sqrt{x^2 + y^2}$, and $\theta = tan^{-1}\left(\frac{y}{x}\right)$.

Depending on the moment order and repetition, the image is encoded in terms of real-valued features that would be used for classification. As the order increases, we get more features which improves the discriminating task among different individuals. In this study, the feature values are computed from $(R_{1,1})$ till the $9^{th}$ order (i.e., $R_{9,9}$).

As mentioned before, the partial iris patterns are first resized such that they are of the same dimension (60 × 60). Then, for each individual there are four partial iris patterns (2 × 2 of the left and right eye) that would be concatenated together to form 120 × 120 iris pattern. For the feature extraction step, Gabor filters along with Zernike moments (ZMs) are used. Accordingly, the main steps of the proposed system are as follows:

  Step-1: Input: left-eye-image, right-eye-image.
  Step-2: Locate the pupil and Iris.
  Step-3: Calculate the average center coordinates of both pupil and iris.
  Step-4: Extract the left and right partial iris patterns for both eyes of the individual.
  Step-5: Concatenate the partial patterns to form 120 × 120 pixel image.
  Step-6: Using 2D-Gabor filters and Zernike moments to extract the features.
  Step-7: Create the feature vector.
  Step-8: Match the extracted feature vector with the iris database.

## 4   Experimental Results

The experimental results provided in this section define the performance of the proposed system. To demonstrate the efficiency of the proposed method, the experiments were conducted using sample images from the following databases:

- **Set 1:** The "CASIA-IrisV4 iris database" [27] of the Institute of Automation, Chinese Academy of Sciences (CASIA). It contains iris images from more than 1,800 individuals. All these images are grayscale images and publicly available. This database is an enhanced version of CASIA-IrisV1, CASIA-IrisV2 and CASIA-IrisV3.
- **Set 2:** Our own iris dataset. This dataset consists of 350 iris images collected from different sources and including various image types.

For Set 1 database (i.e., CASIA-Iris) the system achieves an accuracy of 93.40% while it achieves 89.14% accuracy for Set 2 database as shown in Table 1. The reason for the variability of results with the datasets returns to the fact that Set 1 contains a set of images captured in a constrained environment that makes iris images ideal, whereas Set 2 consists of iris images collected from different sources.

**Table 1.** Experimental results using two datasets

| Iris database | Detection rates | | |
|---|---|---|---|
| | Correct identification | False identification | Accuracy |
| CASIA | 467 | 33 | 93.40% |
| Our database | 312 | 38 | 89.14% |

## 5   Conclusion

In real world imaging systems, the imaging conditions are usually non-ideal; the captured image is a degraded version of the original scene due to different issues such as imaging illumination, shadows, occlusions, lens aberration, wrong focus, motion of the scene, systematic and random sensor errors, etc. In pattern recognition problems, these variations degrade the quality of the training data.

The main contribution of this study is the introduction of a novel iris pattern for human identification; i.e., partial iris pattern, instead of the whole iris. Then, a feature-set consisting of Gabor and ZMs is used for feature extracting and matching. The ZMs features are calculated till 9th order and applied on two datasets. The experiential results show that the system is reliable and promising. For future work, more modalities can be applied to improve the performance for implementing an effective iris biometric system.

# References

1. Li, Y., Xu, X.: Revolutionary information system application in biometrics. In: Proceedings, 2009 International Conference on Networking and Digital Society (2009)
2. Tan, C.-W., Kumar, A.: Accurate iris recognition at a distance using stabilized iris encoding and Zernike moments phase features. IEEE Trans. Image Process. **23**, 3962–3974 (2014)
3. Adini, Y., Moses, Y., Ullman, S.: Face recognition: the problem of compensating for changes in illumination direction. IEEE Trans. Pattern Anal. Mach. Intell. **19**, 721–732 (1997)
4. Xie, X., Lam, K.-M.: Face recognition under varying illumination based on a 2D face shape model. Pattern Recogn. **38**, 221–230 (2005)
5. Kaur, B., Singh, S., Kumar, J.: Iris recognition using Zernike moments and polar harmonic transforms. Arab. J. Sci. Eng. **43**, 7209–7218 (2018). https://doi.org/10.1007/s13369-017-3057-2
6. Tomeo-Reyes, I., Ross, A., Clark, A.D., et al.: A biomechanical approach to iris normalization. In: Proceedings, 2015 International Conference on Biometrics (ICB) (2015)
7. Wildes, R.P.: Iris recognition: an emerging biometric technology. Proc. IEEE **85**, 1348–1363 (1997)
8. Scheuermann, D., Schwiderski-Grosche, S., Struif, B.: Usability of biometrics in relation to electronic signatures, GMD-Forschungszentrum Informations technik Sankt Augustin (2000)
9. Lin, Z., Lu, B.: Iris recognition method based on the optimized Gabor filters. In: Proceedings, 2010 3rd International Congress on Image and Signal Processing (2010)
10. Flom, L., Safir, A.: Iris recognition system, Google Patents (1987)
11. Johnson, R.: Can iris patterns be used to identify people. Los Alamos National Laboratory, CA, Chemical and Laser Sciences Division, Rep LA-12331-PR (1991)
12. Daugman, J.: High confidence personal identification by rapid video analysis of iris texture. In: Proceedings, Proceedings 1992 International Carnahan Conference on Security Technology: Crime Countermeasures (1992)
13. Daugman, J.: High confidence visual recognition of persons by a test of statistical independence. IEEE Trans. Pattern Anal. Mach. Intell. **15**, 1148–1161 (1993)
14. Daugman, J: Biometric personal identification system based on iris analysis, in, USA Patents (1994)
15. Daugman, J.: How iris recognition works. In: The Essential Guide to Image Processing, pp. 715–739. Elsevier (2009)
16. Boles, W.W., Boashash, B.: A human identification technique using images of the iris and wavelet transform. IEEE Trans. Signal Process. **46**, 1185–1188 (1998)
17. Te Chu, C., Ching-Han, C.: High performance iris recognition based on 1-D circular feature extraction (2005)
18. Hosny, K.M., Hafez, M.A.: An algorithm for fast computation of 3D Zernike moments for volumetric images. Math. Probl. Eng. **2012**, 17 (2012)
19. Pirasteh, A., Maghooli, K., Mousavizadeh, S.: Iris recognition using localized Zernike's feature and SVM. Int. Arab. J. Inf. Technol. (IAJIT) **13**, 552–558 (2016)
20. Flusser, J.: Moment invariants in image analysis. In: Proceedings of World Academy of Science, Engineering and Technology (2006)
21. Chen, Z., Sun, S.-K.: A Zernike moment phase-based descriptor for local image representation and matching. IEEE Trans. Image Process. **19**, 205–219 (2009)
22. Rai, H., Yadav, A.: Iris recognition using combined support vector machine and Hamming distance approach. Expert Syst. Appl. **41**, 588–593 (2014)
23. Liu, Y., He, F., Zhu, X., et al.: The improved characteristics of bionic gabor representations by combining with sift key-points for iris recognition. J. Bionic Eng. **12**, 504–517 (2015). https://doi.org/10.1016/S1672-6529(14)60141-4

24. Bhateja, A.K., Sharma, S., Chaudhury, S., et al.: Iris recognition based on sparse representation and k-nearest subspace with genetic algorithm. Pattern Recogn. Lett. **73**, 13–18 (2016)
25. Noh, S.-I., Pae, K., Lee, C., et al.: Multiresolution independent component analysis for its iris identification. In: Proceedings, ITC-CSCC: International Technical Conference on Circuits Systems, Computers and Communications (2002)
26. Ma, L., Wang, Y., Tan, T.: Iris recognition using circular symmetric filters. In: Proceedings, Object Recognition Supported by User Interaction for Service Robots (2002)
27. CASIA-Iris database, Institute of Automation, Chinese Academy of Sciences (2019)
28. MMU-Iris Database, Multimedia-University Malaysia (2016)
29. WVU-Iris Database, West-Virginia-University (2019)
30. Masek, L.: Recognition of human iris patterns for biometric identification. Master's thesis, University of Western Australia (2003)
31. Bakštein, E.: IRIS recognition (2019). (http://www.mlmu.cz/wp-content/uploads/2014/09/Iris-MLMU.pdf)
32. Li, J., Allinson, N.M.: A comprehensive review of current local features for computer vision. Neurocomputing **71**, 1771–1787 (2008)
33. Mikolajczyk, K., Schmid, C.: A performance evaluation of local descriptors. IEEE Trans. Pattern Anal. Mach. Intell. **27**, 1615–1630 (2005)
34. Teague, M.R.: Image analysis via the general theory of moments. JOSA **70**, 920–930 (1980)
35. Hse, H., Newton, A.R.: Sketched symbol recognition using zernike moments, Proceedings. In: Proceedings of the 17th International Conference on Pattern Recognition, ICPR 2004 (2004)

# Analysis of Hollywood's Film Production Using Background Subtraction Visual Activity Index Based on Computer Vision Algorithm

Kristian Dokic[1](✉), Dubravka Mandusic[2], and Lucija Blaskovic[2]

[1] Polytechnic in Pozega, 34000 Pozega, Croatia
kdjokic@vup.hr
[2] Faculty of Agriculture, 10000 Zagreb, Croatia
{simunovic,lmarkic}@agr.hr

**Abstract.** The amount of available video content is increasing every day. Apart from the classic channels of video content distribution, such as television and satellite broadcasting, more and more video content is distributed over the Internet. For some decades, some authors have analysed the properties of the available video content and noted certain trends, and above all, a growing visual activity. This paper describes two measures of visual activity (Visual Activity Index and Average Shot Length), as well as papers that analyse their influence. Also, a new measure based on computer vision algorithms is proposed, and this measure is compared with Visual Activity Index. This proposed measure is more resistant to noise than Visual Activity Index, and this insensitivity is based on selected background subtraction algorithm that has this feature in its base for the proposed measure. The proposed measure called the Background Subtraction Visual Activity Index (BGSVAI) is used at the end of the paper to measure the visual activity of 50 films, which received the Award for Best Picture from the Academy of Motion Picture Arts and Sciences from 1965 to 2014. Also, the visual activity of these films was analysed in the mentioned period.

**Keywords:** Visual activity · Background subtraction algorithms · VAI · BGSVAI

## 1 Introduction

The very notion of "visual activity" has multiple meanings. The key division of areas in which the term is used is the subject's visual activity and the visual activity of the object. The visual activity of the subject primarily deals with a particular branch of medicine, where the visual activity is generally considered to be the activity of eye muscle, and there are various methods of measuring such activity. The second mentioned meaning of visual activity is the visual activity of the object, which means the movement of the object concerning the subject, i.e. the observer. Further in this paper, this second meaning of the term "visual activity" will be used exclusively for dynamic visual media.

Visual activity is encountered in one part of the visual media. The categorisation and a criterion based on which we can determine the existence of visual activity is

© Springer Nature Switzerland AG 2020
A. M. Al-Bakry et al. (Eds.): NTICT 2020, CCIS 1183, pp. 233–245, 2020.
https://doi.org/10.1007/978-3-030-55340-1_17

given by the author, Sutcliffe [1], who divides the media into static and mobile, stating that the borderline speed of image transition above which several images are called moving images - 10 frames per second. For a series of static images, an example of a computer presentation with five images in 60 s is mentioned. The categorisation of the author mentioned above, who has divided the visual media into six categories, has been described while the visual activity is encountered in only the following two:

a)  actual motion picture - video recording
b)  animated motion picture - animated movie, computer game snapshot [1]

Gibson defines the concept of visual activity in the movie as the totality of moving objects and people with a constant background and visual information obtained by moving the observers. The movement of objects and people with a constant background is called a motion, and in the movie, it represents a shot with a still camera in which objects and people move. Movement of the observer is called a movement, and in the movie it is created by moving the camera or lens on the camera [2]. The following definition of the concept of visual activity will be used below.

## 2    Measurement of Visual Activity

If we take the definition of visual activity as suggested by Gibson for the starting point, we can conclude that we are practically all aware of the differences in the visual activity of different types of motion pictures. The visual activity of action movies is generally higher than the visual activity of documentary films. Likewise, it can be generalised and assumed that American animated movies are visually more active than animated Czech film productions from the 1970s. This can be concluded by watching a few animated films from the "Tom and Jerry" series by William Hanna and Joseph Barbera, and "Krtek" (eng.*Mole*) by Zdeněk Miler. However, if you would like to quantify this visual activity, the vast majority of viewers probably would not know how to express it with a number.

There are two approaches that can be found in the literature and serve to quantify the visual activity of motion pictures, whether they are realistic motion pictures (movies, television shows) or animated motion pictures (animated films, computer game recordings).

### 2.1    Measuring Shot Length

The most known method of quantifying motion pictures is the measurement called Average Shot Length (ASL).

Turkovic [3] defines the shot as a "unit of movie exposition, a part of a film in which a spontaneous event is observed without any interruptions." The same author further defines what the shot might consist of, and states that "it may consist of a single shot (shot-image), but usually consists of several, linked in a way that they typically create the impression of apparent movement."

The simplest method to calculate the average shot length is counting shot transitions and dividing movie time with the number of shots in the movie. Salt [4] first mentioned this measure in 1974. He concluded that movies produced by individual authors usually have very similar average shot lengths. The author analysed approximately fifty movies, and besides the mentioned measure, he also studied the distribution of shot-length frequencies that he concluded were closest to Poisson's distribution. Literature analysis related to the average shot length indicates that this measure is most often the focus of a narrow branch of science called Cinemetrics. Baxter [5] for the term Cinemetrics says it is a "statistical analysis of quantitative data, description of the structure and content of movies which could be observed as aspects of style". Brunick et al. [6] suggested later that this measure be renamed into the Average Shot Duration (ASD) because the term Shot Length is also used to measure camera focal length.

The average shot length can be compared in one author's movies, where similarities are usually found, and the difference in the average shot lengths can be seen through film history. The analysis of movies history points to the existence of virtually two periods in which the following regularities are observed:

a)  silent movie period
b)  sound movie period

In the silent movie period, the average shot length was very short for the simple reason that there was no sound in the film, and that was compensated by shots containing the text. With the arrival of sound in the film, the average shot length suddenly increased dramatically, to keep falling continuously after that [7].

DeLong et al. [8] point to a lack of measurements of average shot length, which is caused by the fact that distribution of shot lengths in films does not follow the normal distribution curve, but is rather lognormal distribution. The reason is that most of the shots are short and a few long ones affect the increase in the mean value. Authors conclude that a median is a better measure.

Schaefer et al. [9] used median as a measure of shot length, but in their paper, they analyse shot length in the evening half-hour television news broadcasted by the three major commercial companies, American Broadcasting Company (ABC), Columbia Broadcast System (CBS) and National Broadcasting Company (NBC). The analysis was made for the period between 1969 and 2005, and the pattern, besides the years mentioned above, includes news half-hour television shows from 1983 and 1997. A statistically significant drop in shot length is observed between 1969 and 1983, as well as between 1997 and 2005.

Concerning the assumption that shot lengths were distributed by a lognormal distribution, conflicting attitudes and arguments arose in the scientific community in support and against this assumption. Redfer [10] denies this assumption in his paper, while Baxter [5] pointed to some shortcomings in the methodology used by Redfern. In any case, there is still no consensus regarding the distribution of shot length.

Most of the papers on the average shot length in movies have been written in the last ten years, and one of the reasons is that at www.cinemetrics.lv you can find a database with average shot length values for tens of thousands of movies. The database is constantly updated, and the number of movies processed is constantly increasing. Researchers and enthusiasts from around the world have contributed to the creation of the database, and the results are publicly available to everyone. The database was created at the initiative of the aforementioned author, Yuria Tsiviana, who added a program called CineMetrics on the website, a program that can easily measure the movie's shot length. Without the mentioned program, measuring shot length itself was much slower, and researchers without publicly available databases were forced to work on the movies they analysed themselves.

## 2.2 Visual Activity Index

Visual activity index is mentioned by Bruce [11] and Bertelo [12], but the definition by Cutting et al. [7] will be used in this paper to measure the amount of movement in movies. Authors say that this measure "measures the amount of movement of objects in the shot, as well the movement of a camera itself, in the entire movie". The result obtained by measuring is the number between 0 and 2, with a lower value indicating less movement in the movies.

The measuring visual activity, according to Cutting is based on reduction of each image in a movie into a $256 \times 256$-pixel images. Also, each image is converted to a JPG image file, and the colours are removed, and each point receives the value of a grey colour between 0 and 255. The result of the above-described preparation is approximately 165,000 single pairs of images per movie, and due to hybrid images at NTSC encoding, images separated by a single image are compared in the next step, images 1 and 3, images 2 and 4 etc. until the last picture. The authors calculated Pearson coefficient of correlation for the comparison of the image, with the correlation assuming value 1 in case the two images are identical. In the case of larger visual activity, the images will be less similar, and the correlation will be smaller. The final value of the visual activity index is obtained by subtracting the correlation value from number 1, which gives the theoretical scope for the visual activity index between 0 and 2 [7].

Due to the fairly large similarities between the individual images obtained in the manner described, the authors gained values very close to zero, and the median index of visual activity for 25 million comparative images was 0.034. Distribution of the value of the Visual Activity Index is also important to the authors, indicating that it is not normally distributed.

The authors have processed 145 movies produced between 1935 and 2005 in Hollywood studios and calculated the values of the Visual Activity Index. The value of Visual Activity Index raised during that period. The straight line, i.e. linear function interpolating values, has a value of 0.02 for 1935, while for 2005 this value is 0.06. A threefold increase in value in the observed period of 70 years can be seen [7].

Visual Activity Index presented by the author in 2011 was mentioned several times in various papers, so Salt [13] used it in his paper in which he analysed the pairs of movies that have been filmed with the same scenario but a few decades apart. DeLong et al. [14] used the index to determine the upper limit to which it can be increased with viewers still recognising the content. They used results of Potter [15] who has developed a methodology called "Rapid Serial Visual Presentation" (RSVP) with the aid of which they reached the upper limit of recognising the content of a series of consecutive images, and that limit is 100 ms. In other words, if the viewer is exposed to different images at 10fps, that represents the upper limit at which the image content is still recognisable.

Authors point out that such a series is acceptable to viewers for a very short time, while the average RSVP series of images have a visual activity index of around 0.80.

# 3 Influence of Visual Activity

Visual activity is necessary and expected in dynamic visual media, and Potter has defined the upper limit of visual activity after which the viewer does not register some parts of the content at all. Below is a review of papers that cover the impact of different levels of visual activity on the viewer [15].

## 3.1 Influence of Shot Change

More authors have been studying what happens to viewers after a cut in video content. One of the methods used is called the Secondary Task Response Time (STRT), and it serves to determine the difference in the amount of allocated cognitive resources in different situations. The method consists of measuring the time from the moment the sound is transmitted from the speaker to the spectator response that was previously agreed for that stimulus, and this is typically the push of a button. In this way, the amount of allocated cognitive resources allocated to the primary task, which, in the case of research in this area, paying attention to video content, is measured. The secondary task is the mentioned pressing of the button as a reaction to audio stimuli. As the viewer is more focused on video content, his/her reaction after the sound signal is slower, suggesting that more cognitive resources are assigned to the primary task. Geiger [16] states that the reaction time after the cut in video content is longer compared to the reaction time before the cut, suggesting that after the cut multiple cognitive resources are allocated for the analysis of the new shot. In addition, the author has found that in the case where the two shots are unconnected, the content viewer allocates more cognitive resources in order to link the information from the new shot with the information from the previous one, whereas in the case when the cut divides shots with linked content, the amount of allocated resources is lower. Carroll and Bever [17] confirm the allocation of more viewers' cognitive resources after the cut between two shots with their research that showed that video content is faster recognisable than when they are after the cut than when they are directly before the cut.

There are several papers describing physiological changes in viewers after cuts in video content. Reeves et al. [18] indicated that half a second after the cut, and up to 1.5 s after the cut, the amount of brain wave wavelengths measured by the electroencephalograph (EEG) drops, but rises again after 2 s. The authors state that this is consistent with the literature in that field where it is stated that following the introduction of a new stimulation there is a sharp drop in the amount of alpha waves, and after that, a gradual recovery to the previous level. Authors link this phenomenon to orienting response, and in the field of media, changes in attention occur as a result of the structural features of a television program [19, 20]. Among other things, the structural features of a television program include cuts between shots. A psychological and physiological reaction called orienting response is not closely related to the media, and can occur at any unexpected sensory stimulation [21].

Lang [22] has studied the relationship between the heart rate and circumstances under which the aforementioned change of attention appears, indicating the existence of a link between the structural features of a television program and heart rate in a way that after a transition between shorts there is a slight decrease in heart rate. The orienting response is usually measured in heart rate and lasts for ten beats, averaging 8.33 s, suggesting that not every cut can cause this orienting response because cuts are often more frequents in the media [23].

### 3.2 Impact of Changing ASL Values

Video content with different ASL values was used by Chandler and Pronin [24], with three three-minute video clips having ASL values: 0.75 s, 1.5 s and 3 s. Video clips are shown to participants in the experiment previously divided into three groups, and the authors confirmed their hypothesis that the participants who watched the video with the smallest ASL value stated that they thought the fastest, followed by a group with middle and with the highest value of ASL. Also, the results of Cognitive Assessment of Risky Events (CARE) tests that use prediction of risk behaviour in the future have indicated that participants who viewed the video with the smallest ASL values have a greater likelihood of risk behaviour in the next six months. As the value of ASL increased, the likelihood of risky behaviour was also lower. Participants exposed to lower ASL values had a lower level of perception that risk behaviour has negative consequences, and ASL value growth has also increased the level of perception. The authors concluded that practically everyone should be aware of this phenomenon, especially people in responsible and managerial positions. They also point to the current situation in which parents and legislators are sometimes concerned about the unacceptable content of the media children are exposed to (violence, erotica) and confirm that they are right but that the problem is obviously not just in the content [25].

Pronin and Wegner [26] used PowerPoint presentations in their research, which featured two types of statements in two speeds, and in that way, they tested the impact of speed on mood. The statements consisted of a series of letters that appeared at speeds of 40 ms or 170 ms, while the average reading speed is approximately 80 ms per letter. After the entire statement was shown on the screen, the presentation would go to the next slide with a pause of 320 ms or 4000 ms. Authors proved in their paper the hypothesis that there is a connection between the speed of thinking and the positive mood, regardless of the content of statements presented. In the study, the participants who were divided into two groups were presented with statements of depressive and stimulating content, but this did not affect the positive mood of the group to which the statements were shown at a higher speed. While the term ASL is not mentioned in the paper, it is obvious that authors used it.

### 3.3 Influence of Different VAI Values

The Visual Activity Index was presented to the scientific community in 2011, so it is hard to expect that it is currently being accepted and used in research. However, Candan et al. [27] used VAI in their research on memory skills concerning the form of displayed content, using video clips with shot transitions and static image with transitions. Authors have pointed out that the content displayed in the form of video clips is better remembered, and the value of visual activity index of individual video clips does not affect the memory. The authors conclude that the presence of movement itself, rather than the amount of the same, affect a better memory.

## 4  More Robust Measure of Visual Activity - BGSVAI

Measures described above have certain drawbacks. The first measure mentioned, the Average Shot Length, completely neglects the visual activity that occurs between the two cuts. For this reason, it is often used in the analysis of individual authors' films to determine a change of style over time.

The second measure mentioned, Visual Activity Index, is much more complex and measures any visual activity within a video clip regardless of the number of cuts. However, the key disadvantage of this measure is the sensitivity to noise. Noise is the degradation of the image quality that arises for several reasons. It can arise on the camera sensor itself, which converts the electromagnetic radiation of the visible spectrum into voltage, but also when transmitting and processing the signal.

For more accurate measurement of visual activity, computer vision algorithms can beused. They are usually designed to neglect certain types of noise in a video while performing certain processing. One of the subcategories of computer vision algorithms is the background subtraction algorithms. Jain [28] proposed one of the first algorithms for removing the background, and today there are dozens of them. The algorithms handle the video content in such a way that output is a two-colour video in which one colour indicates a stationary background while the other colour indicates a moving object. They are usually black and white. The basic idea of the new way of measuring visual activity is that the ratio of a number of pixels of the moving object and the total number of pixels gives the measure of the visual activity of a particular frame in the video. If we calculate this ratio for the entire video, we get the measure of visual activity. This can be written in the form of:

$$BGSVAI = \Sigma \, n_{fg} / \Sigma \, n_{all} \tag{1}$$

In the formula, $n_{fg}$ is the number of pixels of moving objects, while $n_{all}$ is the total number of pixels. BGSVAI acronym comes from the Background Subtraction Visual Activity Index. Selection of the algorithm from the above-mentioned dozens of algorithms is described in more detail by Dokic et al. [29]. The algorithm called *Multi-Layer Background Subtraction Based on Colour and Texture* was selected, and its main task is to change moving objects pixels from black to white colour [30].

To test the proposed measure for visual activity, it can be compared to the existing and known measure - Visual Activity Index. The hypothesis can be set that a statistically significant association of values obtained with two different measures is expected: VAI and proposed BGSVAI.

To test the hypothesis, VAI and BGSVAI values were calculated for 135 video clips ranging in length between 60 to 120 s. To calculate the value, the original algorithm in MATLAB for VAI by the author Cutting was used, while a background Multilayer removal algorithm was used to calculate BGSVAI. Multilayer removal algorithm was implemented as an additional library in the OpenCV program library that was used in the program written by C++ programming language in the Microsoft development environment Visual C++ Express 2010. For correlation testing, Spearman's correlation coefficient was calculated, which is:

$r_s = 0.823$,

while Pearson's correlation coefficient is:

$r = 0.793$

In both cases $p < 0.01$, so correlation values are statistically significant. This is a strong positive correlation according to Horvat [30] since the correlation coefficient is between 0.7 and 1.

Figure 1 shows graphically the correlation between VAI and BGSVAI values for video clips at the scattered graph.

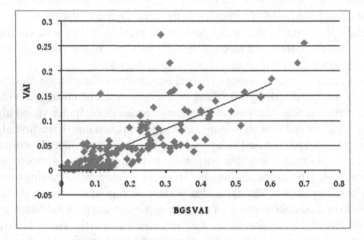

**Fig. 1.** Correlation between VAI and BGSVAI

The strong correlation suggests that VA and BGSVAI are likely to measure the same phenomenon - visual activity.

## 5  Analysis of Hollywood's Film Production

ASL and VAI measures were used by authors for the analysis of American film production and, as noted above, pointed to the growth of visual activity and the constant reduction of shot length. For testing BGSVAI, a logical choice is American film productions because the obtained values and results can be compared to the results obtained with ASL and VAI measures.

Random selection of movies, unfortunately, has not been taken into account for several reasons, and the key reason is that access to all American film industry films produced over the last fifty years was extremely expensive. For this reason, the sample for analysis is appropriate, and it consists of fifty movies that received the Academy Award for Best Picture between 1965 and 2014 from the Academy of Motion Picture

Arts and Sciences. Tables with processed movies and their characteristics can be found on the site: http://kristiandokic.from.hr/.

Since there are fifty movies that have about two hundred thousand individual frames each, over 10 million individual images in total have been processed. The algorithm was tested on several computers, and the processing time for each image was between 2 and 5 s. On newer and faster PCs with a modern processor, this time was 2 s, making the total processing of all 50 movies a little over seven months. For this reason, movie processing is distributed to 11 weak computers with AMD processor. To avoid configuring each computer, a *VMware Player* program was used. The configuration of the environment is made only once and the image file of the complete hard disk is copied to other computers. On one used computer, processing of all movies would last approximately a bit less than a year and considering that 11 computers were used at the same time, processing took just over a month. The processing time for one image was approximately 3 s. During the processing, it was necessary to monitor the computers and after finishing processing one movie, it was necessary to start processing the next one. Movie processing duration was between 5 and 9 days, depending on the resolution and duration of the movie itself.

In Fig. 2, the obtained BGSVAI values are presented, and on the horizontal axis are the years in which the movies received Academy Award for Best Picture. The graph shows two anomalies, which are significantly lower BGSVAI values for two films, from 1993 and 2011. These are the movies "Schindler's List" and "The Artist" that are produced in black and white, which obviously has an impact on the result. Because of those other candidates from the same years were selected. They were "The Fugitive" directed by Andrew Davis and "War Horse" directed by Steven Spielberg.

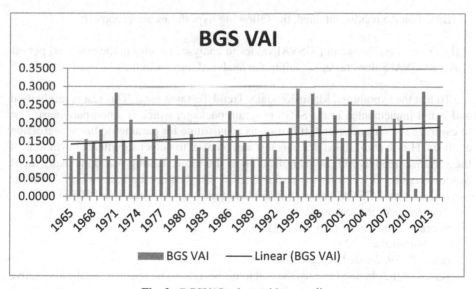

**Fig. 2.** BGSVAI values with anomalies

In Fig. 3, BGSVAI values after movies "Schindler's List" and "The Artist" have been replaced, as can be seen. The anomalies associated with the use of black and white recording techniques is no longer visible.

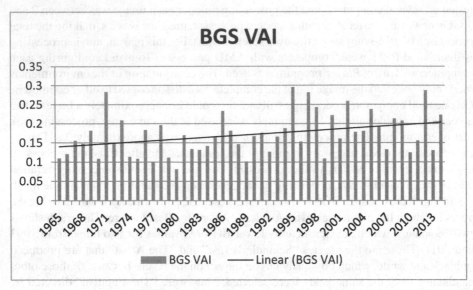

**Fig. 3.** BGSVAI values without anomalies

Based on the results obtained, the following hypotheses are proposed:

H.0 - There is no change in BGSVAI values in analysed movies in the observed period
H.A - BGSVAI values raise, i.e. fall in the analysed movies in the observed period

To test the hypothesis, Mann-Kendal's Trend Test was used. This test is rarely used and is not implemented in the SPSS program package, which is a non-parametric test to estimate the variance trend. It is used as a substitute for parametric linear regression analysis. Hirsch et al. stated that this test can be used to detect significant deviations of the variance trend as well as quantification of the same. Testing was done using the R Studio version 1.0.153 software package. BGSVAI data can be seen in Fig. 4. On the vertical axis are BGSVAI values, but displayed as ranks between 1 and 50 [31–35].
The results of the test are:
Score = 293, Var (Score) = 14291.67
Denominator = 1225
tau = 0.239, 2-sided $p_{value}$ = 0.014584
Based on the obtained results, the null hypothesis must be rejected and the alternative one must be accepted at a significance level of 5%, and considering that the S > 0, the trend is growing. Thus, we can conclude:

H.A - BGSVAI values grow in the analysed movies in the observed period

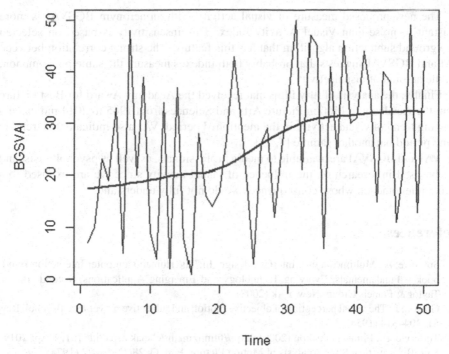

**Fig. 4.** BGSVAI as ranks

## 6 Discussion and Conclusion

Many authors, especially in the fields of science that deal with children and youth, warn of a steady increase of time spent watching at some kind of a screen. With the development of technology, the content on these screens is increasingly dynamic. As noted above, some authors have been analysing the available video content for a long time and have noticed the growing trend of visual activity. Usually, the authors use two measures, namely VIA and ASL. Furthermore, research has shown that changes in VAI and ASL have different physiological and physiological effects on viewers. Nevertheless, despite this constant growth of visual activity, state institutions around the world have no control over these phenomena.

Average Shot Duration (ASD) is the most used method, and it is described in this paper. The main disadvantage of ASD is that it can measure only one movie feature connected with visual activity. Proposed BGSVAI can measure all features that are connected with the visual activity because it measures difference between every two frames in a movie.

The second one and the last one method for visual activity measuring is Visual Activity Index that is described in this paper, too. Described VAI method is much more advanced than ASD but the main VAI failing is that is sensitive to noise in video. Noise is registered as an activity by VAI. A source of noise can be imperfection of the camera sensor, as well as a swinging leaf in wind in the video.

The new proposed measure of visual activity with an acronym BGSVAI is more resistant to noise than Visual Activity Index. This insensitivity is based on selected background subtraction algorithm that has this feature. The strong correlation between VAI and BGSVAI suggests that probably both indexes measure the same phenomenon, i.e. the amount of visualactivity.

Finally, the analysis of fifty films that received the Academy Award for Best Picture from the Academy of Motion Picture Arts and Sciences from 1965 to 2014 indicated a growing trend of visual activity in the mentioned period. VAI also indicate this trend in some period but another sample [7, 13].

VAI and BGSVAI are available to media professionals, as well as psychologists, and can be used in research on the influence of visual activity since we are exposed to a global phenomenon whose consequences we do not fully understand.

# References

1. Sutcliffe, A.: Multimedia user interface design. In: The Human-Computer Interaction Handbook – Fundamentals, Evolving Technology and Emerging Applications, Second edition. Taylor & Francis Group, New York (2008)
2. Gibson, J.: The visual perception of objective motion and subjective movement. Psychol. Rev. **61**, 304–314 (1954)
3. Turkovic, H.: Filmski leksikon (2008). http://film.lzmk.hr/clanak.aspx?id=761. 1 Apr 2019
4. Salt, B.: Statistical Style Analysis of Motion Picture. Film Q. **28**(1), 13–22 (1974)
5. Baxter, M.: Notes on Cinemetric Data Analysis (2014). http://www.cinemetrics.lv/dev/Cin emetrics_Book_Baxter.pdf. Accessed 14 Apr 2019
6. Brunick, K., Cutting, J., DeLong, J.: Low-level features of film: what they are and why we would be lost without them. In: Psychocinematics – Exploring Cognition at the Movies (2013)
7. Cutting, J., DeLong, K., Brunick, L.: Visual activity in Hollywood film: 1935 to 2005 and beyond. Psychol. Aesthet. Creat. Arts **5**, 115–125 (2011)
8. DeLong, J., Brunick, K., Cutting, J.: Film through the human visual system: finding patterns and limits. In: The Social Science of Cinema, 1st edn. (2013)
9. Schaefer, R., Martinez, T., Martinez, B., Martinez, M.: Trends in network news editing strategies from 1969 through 2005. J. Broadcast. Electron. Media **53**(3), 347–364 (2009)
10. Redfern, N.: The lognormal distribution is not an appropriate parametric model for shot length distributions of Hollywood films. Literary Linguist. Comput. **30**, 137–151 (2012)
11. Bruce, C.: Primate frontal eye fields. I. Single neurons discharging before saccades. J. Neurophysiol. **53**, 603–635 (1985)
12. Bertolo, H., Paiva, T., Pessoa, L., Mestre, T., Marques, R., Santos, R.: Visual dream content, graphical representation and EEG alpha activity in congenitally blind subjects. Cogn. Brain. Res. **15**(3), 277–284 (2003)
13. Salt, B.: The exact remake: a statistical style analysis of six Hollywood films. New Rev. Film Telev. Stud. **14**, 1–20 (2016)
14. DeLong, J., Brunick, K., Cutting, J.: Film through the human visual system: finding patterns and limits. In: The Social Science of Cinema, 1st edn. (2013)
15. Potter, M.: Short-term conceptual memory for pictures. J. Exp. Psychol. Hum. Learn. Mem. **2**, 509–522 (1976)
16. Geiger, S., Reeves, B.: The effects of scene changes and semantic relatedness on attention to television. Commun. Res. **20**, 155–175 (1993)
17. Carrol, S., Bever, T.: Segmentation in cinema perception. Science **191**, 1053–1055 (1976)

18. Reeves, B.T.E., Rothschild, M.L., McDonald, D., Hirsch, J., Goldstein, R.: Attention to television: intrastimulus effects of movement and scene changes on alpha variation over time. Int. J. Neurosci. **27**, 241–255 (1985)

19. Rothschild, M., Thorson, E., Reeves, B., Hirsch, J., Goldstein, R.: EEG activity and the processing of television commercials. Commun. Res. **13**, 182–220 (1986)

20. Anderson, D., Levin, S., Lorch, E.: The effects of TV program pacing on the behavior of preschool children. AV Commun. Rev. **25**, 159–166 (1977)

21. Lynn, R.: Attention, Arousal and the Orientation Reaction. Pergamon Press Ltd., London (1966)

22. Lang, A.: Involuntary attention and physiological arousal evoked by structural features and emotional content in TV commercials. Commun. Res. **17**, 275–299 (1990)

23. Smith, T.: An attentional theory of continuity editing. University of Edinburgh, Edinburgh (2005)

24. Chandler, J., Pronin, E.: Fast thought speed induces risk taking. Psychol. Sci. **23**, 370–374 (2012)

25. Ketz, E., Fromme, K., D'Amico, E.: Effects of outcome expectancies and personality on young adults' illicit drug use, heavy drinking, and risky sexual behavior. Cogn. Ther. Res. **24**, 1–22 (2000). https://doi.org/10.1023/A:1005460107337

26. Pronin, E., Wegner, D.: Manic thinking: independent effects of thought speed and thought content on mood. Psychol. Sci. **17**, 807–813 (2006)

27. Candan, A., Cutting, J., DeLong, J.: RSVP at the movies: dynamic images are remembered better than static images when resources are limited. Vis. Cogn. **23**, 1205–1216 (2016)

28. Jain, R., Nagel, H.: On the analysis of accumulative difference pictures from image sequences of real world scenes. IEEE Trans. Pattern Anal. Mach. Intell. **2**, 206–214 (1979)

29. Dokic, K., Mandusic, D., Idlbek, R.: Model of information density measuring in e-learning videos. In: 6th Cyprus International Conference on Educational Research (CYICER-2017), Kyrenia (2017)

30. Yao, J., Odobez, J.: Multi-layer background subtraction based on color and texture. In: IEEE The CVPR Visual Surveillance Workshop, Minneapolis (2007)

31. Horvat, J., Mioc, J.: Osnove statistike. Naklada Ljevak d.o.o, Zagreb (2012)

32. Mann, H.: Nonparametric tests against trend. Econometrica **13**(3), 245–259 (1945)

33. Kendall, M.: Rank Correlation Methods, 4th edn. Charles Griffin, London (1975)

34. Gilbert, R.: Statistical Methods for Environmental Pollution Monitoring. Wiley, New York (1987)

35. Hirsch, R., Slacj, J., Smith, R.: Techniques of trend analysis for monthly water quality data. Water Resour. Res. **18**, 107–121 (1982)

# Breast Cancer Recognition by Computer Aided Based on Improved Fuzzy c-Mean and ANN

Zahraa Faisal and Nidhal K. El Abbadi$^{(\boxtimes)}$ (iD)

Computer Science Department, Education College, University of Kufa, Najaf, Iraq
{zahraaf.shouman,nidhal.elabbadi}@uokufa.edu.iq

**Abstract.** Breast cancer is the most common reason of women's disease and also the major reason of cancer deaths around the world. The best test used for examination and early diagnosis is X-ray mammography. The malignant and benign tissues usually appear with weak contrast and often not very clear in mammography image. The proposed method is computer-aided diagnosis to support radiologists in their diagnosis. The proposed algorithm focus initially on detecting tumor in mammograms image based on improved fuzzy c-mean algorithm (FCM). Discrete wavelet transformation (DWT) and principle component analysis (PCA) are implemented on the binary image resulted from detection process prior to extract two types of features which uses as input to the neural network for classification the tumor to benign or malignant. The proposed algorithm was tested over several images taken from the Mammographic Image Analysis Society (MIAS) database. The accuracy of detection the tumor in image is 100%, while the accuracy of classification of tumor was 92.9%.

**Keywords:** Image processing · Mammography · DWT · Cancer · Improved FCM · Breast

## 1 Introduction

Breast cancer appears when cells in the breast begin to grow abnormal and out of control. Breast cancer is the most common cause of death in the women and the second leading cause of cancer deaths worldwide (after lung cancer) [1].

In 2015, the Canadian Cancer Society's statistics for breast cancer report: there are about 25,200 cases are registered to be infected by breast cancer, distributed to (25000) cases in females and (200) cases in males; it is the 2nd most frequent cancer in Canada [2]. In 2010, the American Cancer Society's statistics for breast cancer report: there are about 40,230 cases deaths, distributed to (39,840) cases in females and (390) cases in males [3]. According to Iraqi cancer registry report of (2011) breast cancer is the first most common cancer in Iraq, there are 3845 cancerous cases registered infected by breast, distributed as 82 cases in males and 3763 cases in females [4]. One of the main prevention measures to reduce the effect of breast cancer is early detection. The elementary prevention in early stages of the disease still complex as the reason remain almost unknown. To detect the breast cancer, some typical signs of this disease can

© Springer Nature Switzerland AG 2020
A. M. Al-Bakry et al. (Eds.): NTICT 2020, CCIS 1183, pp. 246–257, 2020.
https://doi.org/10.1007/978-3-030-55340-1_18

be targeted, such as microcalcifications and masses shown on mammograms, which detect by physician by using the naked eye. This process suffers from high percent of error. Therefore, the need to accurate and speed method to diagnose the breast cancer emerged, and that become goal for many researcher, especially in the field of digital image processing. Where, detection of breast cancer in its early phases at large extent reduces death-rate. It also saves hundreds of millions of dollars that would expend on the treatment of disease in developed cases.

## 2  Literature Survey

(Gayathri et al. 2013) The researchers provided a summary to survey about breast cancer diagnosis using various machine learning algorithms and methods, which are used to improve the accuracy of prophesying cancer. This survey can also help us to know about number of researches that are implemented to diagnose the breast cancer [5].

(Khedr et al. 2014) suggested detecting the breast cancer from medical image using traditional Fuzzy Clustering Means (FCM), features for mammogram images are extracted to use in training the classifier to distinguish tumors with good accuracy [6].

(Salahat et al. 2016) presented a method to forming a series of binary threshold intensity images from an MRI image of a patient. Each of the binary threshold intensity image is based on a respective intensity. The binary threshold intensity images are processed to identify one or more bright external regions in which image pixels have the same value, and have a higher intensity than surrounding region pixels in the MRI image. This region may be identified as potentially cancerous [7].

(Prannoy Giri et al. 2017) proposed an analysis method used for texture features for detection of cancer. These texture features are extracted from the region of interest of the mammogram to describe the micro calcifications into harmless, ordinary. These features are passed through Back Propagation algorithm for better understanding of the cancer pattern in the mammography image [8].

## 3  Methodology

### 3.1  DWT Technique

Discrete wavelet transform (DWT) use in image processing to convert images from the spatial domain into the frequency domain (transform image pixels to wavelet). The time domain signal passes in two filters (low and high pass filters) to get two versions (low (L) and high (H) version), each of these versions passes on (low and high pass filters) again. The final result is four sub-band image (LL, LH, HH, and HL). The LL sub-band can be as the approximation component of the image, while the LH, HL, HH sub-bands can be as the detailed components of the image [9].

## 3.2 PCA Technique

Principle component analysis (PCA) is a statistical technique uses orthogonal transformation to convert correlated variables (entities with various numerical values) into a set of values linearly uncorrelated called principle components [10]. In the current research the PCA used to remove the similar value and preserve the values that gives high variance.

## 3.3 Proposed Method

The main steps of Breast Tumor Identification and Recognition show in Fig. 1. The Proposed Method can be divided to the follows steps:

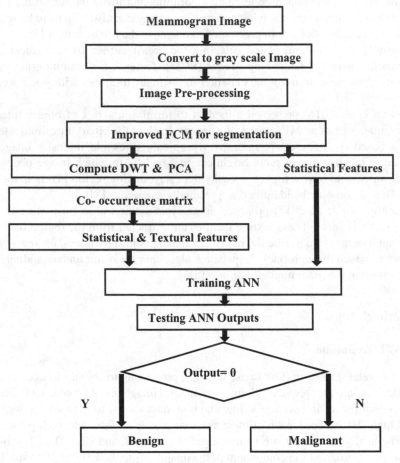

**Fig. 1.** Flow chart for proposed method

## Step One: Pre-processing

Medical images are either color images or gray images, in the current proposal we deal with gray images, for that the RGB images will convert to gray image. Medical images almost have unwanted particles (noise) produced due to different reasons. Noise will affect the diagnostic value.

De-noising techniques play a pivotal role in the enhancement of x-ray image quality. The rank filter suggested as de-noising for the repression of noise from x-ray breast image. Rank filter is robust to eliminates unwanted data while its preserve the edges. The filter orders the neighborhoods pixels' value of the target pixel (pixel at the center of the filter), rank them in ascending order, then the middle value will replace the value of target pixel. Figure 2 show the result of image de-noising.

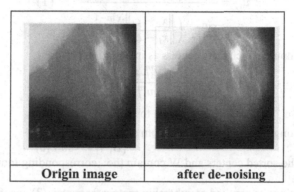

| Origin image | after de-noising |

**Fig. 2.** Shows the results of first step.

**Step Two: Detection Process**
The most important stage in the automatic breast tumor recognition is the detection. Tumor detection is difficult process, due to: weak contrast between the tumor and the surrounding region in image. The proposed detection process based on the following points:

i.   The proposed algorithm to detect the breast tumor depend on improvement the traditional Fuzzy cmean algorithm (FCM). The detection algorithm converts the gray image to binary image, where the value one assigned to tumor, and zero to the rest of breast tissues, as shown in Fig. 3 image B. The proposed algorithm (Algorithm 1) summarized the improved FCM steps, which uses to detect the tumor in the breast image.

---

**Algorithm 1: Improve FCM method**

**Input:** Mammogram Image after Pre-processing
**Output:** binary image.
(1): Convert image to vector.

(2): Choose 3 random centroid in the image.

(3): Calculate membership matrix as in equation (1).
Where c cluster's Number, m > 1.

$$u_{ij} = \frac{1}{\sum_{k=1}^{C} \left( \frac{\|x_l - c_j\|}{\|x_i - c_k\|} \right)^{\frac{2}{m-1}}} \qquad (1)$$

(4): Compute the clusters centers as in equation (2)

$$c_j = \frac{\sum_{j=1}^{C} u_{ij}^m \cdot x_i}{\sum_{j=1}^{C} u_{ij}^m} \qquad (2)$$

(5): if $\left\{ \left| u_{ij}^{(k+1)} - u_{ij}^{(k)} \right| \right\} \geq \varepsilon$, return to Step (3).
(6): Sorting the resulted centers from step (4)

$$\{B, W\} = \text{sort} (c)$$

B lists the sorted centers and W contain the  corresponding indices of c

(7): Sorting the resulted membership matrix from step (3) according to
vector w resulted from Step (6)

$$bw(i, j) = \begin{cases} 1 & u(i, j) \geq K \\ 0 & \text{otherwise} \end{cases}$$

(8): Find the **largest** element in each row and its **index** of resulted
membership matrix.

(9): Compute values of c1, c2 according to index value from step (7)

$$C_i = \begin{cases} c1 & \text{index} = 2 \\ c2 & \text{index} = 3 \end{cases}$$

(10): Compute the value (K) by equation

$$K = \frac{(\max(\text{vector}(c1)) + \min(\text{vector}(c2)))}{2} \qquad (3)$$

(11): Apply the value (K) on mammogram image to obtain a binary image:

$$bw(i, j) = \begin{cases} 1 & u(i, j) \geq K \\ 0 & \text{otherwise} \end{cases}$$

---

ii.  At this step we will enhance the detected tumor image by using mathematical morphology such as (erosion, dilation, and close) to delete the unwanted objects that connected to the image edge, split apart joined objects, fill the small holes, removing the small objects, and repair intrusions. Result show in Fig. 3 image C.

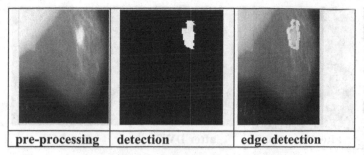

| pre-processing | detection | edge detection |

**Fig. 3.** Phases of detection process of breast tumor.

### Step Three: Features Extraction

Features extraction is very important process which provide the system with useful information and details about the tumor. The features extraction process based on two types:

i.  Statistical Features Extracted [11]:
    Statistical features extracted from the image results from step 1, the features used in this proposal are:

    A.  Mean (M): The average of image, is calculated by Eq. 4.

    $$M = \left( \frac{1}{mn} \sum_{x=0}^{m-1} \sum_{y=0}^{n-1} f(x, y) \right) \tag{4}$$

    B.  Skewness (Ske): is a measure of symmetry or the lack of symmetry. As in Eq. (5).

    $$Ske = \left( \left( \frac{1}{mn} \right) \frac{\sum (f(x, y) - M)^3}{\sigma^3} \right) \tag{5}$$

    C.  Kurtosis (Kur): is a measure of the random variable's probability distribution. As in Eq. (6).

    $$Kur = \left( \left( \frac{1}{mn} \right) \frac{\sum (f(x, y) - M)^4}{\sigma^4} \right) \tag{6}$$

    D.  Major axis Length (mx): its gives a scalar value that specifies the length of the major axis of the ellipse (in pixels).
    E.  Minor Axis Length (mn): its gives a scalar value that specifies the length of the minor axis of the ellipse (in pixels).

ii. Texture Features Extracted by using GLCM Texture analysis give power distinction between malignant and benign tissues, which may not be visible to human eye. So that it's efficient for early tumor diagnosis. To extract these features we have at the first compute the DWT for the image resulted from detection, DWT computes the approximation coefficients matrices (LL, LH, HL, and HH). Then, the PCA applied on the DWT image. The result of this process show in Fig. 4.

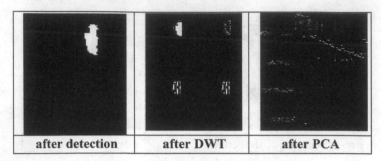

| after detection | after DWT | after PCA |

**Fig. 4.** Image after DWT and PCA

Now, co-occurrence matrix will be built from PCA image. The following textual features are extracted from co-occurrence matrix which derived from PCA image. These features are [11]:

A. Contrast (Con.): is a measure of intensity of a pixel and its neighbor on the image, as in equation.

$$Con = -\sum_{x=0}^{m-1}\sum_{y=0}^{n-1}(x,y)^2 f(x,y) \tag{7}$$

B. Energy (Ene.): defined as the quantifiable amount for repetitive pixel pairs. (Ene.) is a parameter to measure the similarity of an image. As in equation

$$Ene = -\sum_{x=0}^{m-1}\sum_{y=0}^{n-1}f^2(x,y) \tag{8}$$

C. Correlation (Corr.): is the measurement of spatial features dependencies between the pixels. Measured by equation.

$$Cor = \frac{\sum_{x=0}^{m-1}\sum_{y=0}^{n-1}(x,y)f(x,y) - M_x M_y}{\sigma_x \sigma_y} \tag{9}$$

**Step Four: Recognition by Using Artificial Neural Network (ANN)**
Due to the non-linear processing power of ANN's neurons, so that its able to solve very complex problems. The ANN is used for classification of breast tumor into benign or malignant. Machine learning consist of 3 layers: input, hidden, and output layer. In this proposal Back propagation neural network are used for classification of breast cancer. Network show in Fig. 5.

The input for ANN will be the features extracted from image in the previous step. The network after training it will be able to classified the tumor to two types, benign or malignant. Sample of network input show in Fig. 6.

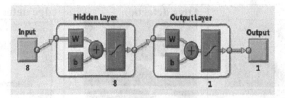

**Fig. 5.** ANN structure

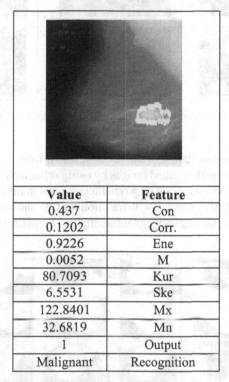

| Value | Feature |
|-----------|-------------|
| 0.437 | Con |
| 0.1202 | Corr. |
| 0.9226 | Ene |
| 0.0052 | M |
| 80.7093 | Kur |
| 6.5531 | Ske |
| 122.8401 | Mx |
| 32.6819 | Mn |
| 1 | Output |
| Malignant | Recognition |

**Fig. 6.** Shows the recognition result of the breast tumor with parameters.

## 4   The Results and Discussion

The proposed algorithm implemented on images provided from the MIAS database, dataset website is (http://netpbm.sourceforge.net/doc/pgm.html). The MIAS database content 115 image, 64 image for benign tumor, and 51 image for malignant tumor. Performance examined of suggested algorithm as follow:

We first test the detection algorithm, this algorithm works efficiently, where the tumor is determined and all the other objects or structures of the breast are removed. Detection accuracy for this method was heartening and reaches up to 100%. where all tumor in images detected and segmented correctly, some of images shows in Fig. 7.

| | After detection | edge detection |
|---|---|---|

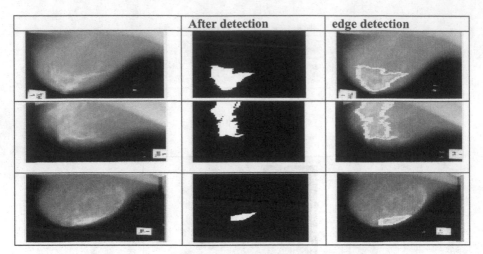

**Fig. 7.** Shows the results of the suggested tumor detection algorithm.

The detection and segmentation of the tumor resulted from the suggested detection algorithm was compared with resulted images by using other techniques. The suggested algorithm gave almost more accurate than other techniques in tumor detection as sow in Fig. 8. The improved FCM compared with traditional FCM and we found that improved FCM more efficient in clustering and segmentation, the comparing results show in Fig. 8.

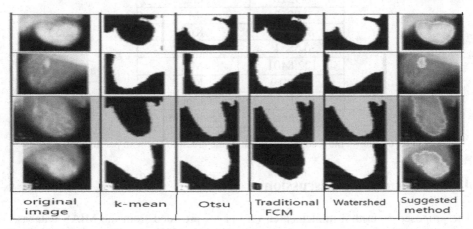

| original image | k-mean | Otsu | Traditional FCM | Watershed | Suggested method |
|---|---|---|---|---|---|

**Fig. 8.** Shows comparing the breast tumor detection between suggested algorithm and other techniques.

In this proposal we suggested to use ANN to classify the tumor. We train the network with 30 mammogram images, and used 85 mammogram images for testing. The recognition accuracy of the suggested algorithm determined by using the Eq. 10. The result of NN classification summarized in Table 1.

**Table 1.** Classification results of ANN.

|  | Classified as benign | Classified as malignant |
|---|---|---|
| Benign (49) | TN = 45 | FN = 2 |
| Malignant (36) | FP = 4 | TP = 34 |

$$Accuracy = \frac{TP + TN}{TP + FP + TN + FN} \tag{10}$$

The accuracy reaches nearly 92.9% according to Eq. (10). Figure 9 show sample of benign and malignant tissues and their features.

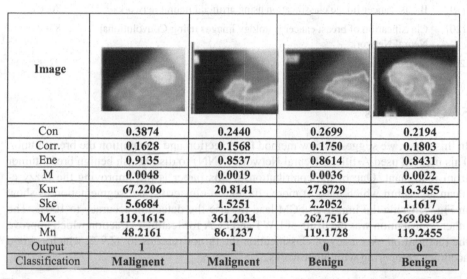

| Image | | | | |
|---|---|---|---|---|
| Con | 0.3874 | 0.2440 | 0.2699 | 0.2194 |
| Corr. | 0.1628 | 0.1568 | 0.1750 | 0.1803 |
| Ene | 0.9135 | 0.8537 | 0.8614 | 0.8431 |
| M | 0.0048 | 0.0019 | 0.0036 | 0.0022 |
| Kur | 67.2206 | 20.8141 | 27.8729 | 16.3455 |
| Ske | 5.6684 | 1.5251 | 2.2052 | 1.1617 |
| Mx | 119.1615 | 361.2034 | 262.7516 | 269.0849 |
| Mn | 48.2161 | 86.1237 | 119.1728 | 119.2455 |
| Output | 1 | 1 | 0 | 0 |
| Classification | **Malignent** | **Malignent** | **Benign** | **Benign** |

**Fig. 9.** Shows samples of the recognition results of the breast tumor with parameters for suggested algorithm.

The performance of the suggested method is compared with other methods, the results listed in the Table 2.

**Table 2.** The comparison results of the suggested method with the like works.

| Ref. | Method name & source | Precision |
|------|---------------------|-----------|
| [14] | Use of the bootstrap technique with small training sets for computer-aided diagnosis in breast ultrasound | 87.07% |
| [15] | Computer-aided diagnosis of solid breast nodules: use of an artificial neural network based on multiple sonographic features | 91% |
| [16] | Feature selection and classification of breast cancer on dynamic Magnetic Resonance Imaging by using artificial neural networks | 91% |
| [17] | Feature selection and classification of breast cancer on dynamic Magnetic Resonance Imaging using ANN and SVM | 86% |
| [18] | Mammogram analysis using feed-forward back propagation and cascade-forward back propagation Artificial Neural Network | 87.5% |
| [19] | Breast cancer image classification using artificial neural networks | 70.4% |
| [20] | Classification of breast cancer histology images using Convolutional Neural Network | 83.3% |
| 2019 | Suggested method | 92.9% |

## 5  Conclusions

In this paper, we suggested new method for detection and recognition the breast tumor. This method used Artificial Neural Network (ANN) to distinguish benign breast tumor from malignant. One of the contribution of this paper is the improving the fuzzy c-mean algorithm which make it more efficient in clustering and segmentation as show in Fig. 7. Using the improved FCM lead to raise the detection accuracy to 100%. The other contribution was the using of DWT and PCA for feature extraction which improve the recognition. Accuracy of classification the tumor reaches to 92.9%, which is good accuracy comparing with other works.

## References

1. Deserno, T.M.: Fundamentals of biomedical image processing. In: Deserno, T. (ed.) Biomedical Image Processing. Biological and Medical Physics, Biomedical Engineering. Springer, Heidelberg (2010). https://doi.org/10.1007/978-3-642-15816-2_1
2. Statistics, C.: Canadian cancer society's advisory committee on cancer statistics. in., Canadian Cancer Society, Canadian Cancer Society (2015)
3. Jemal, A., et al.: Cancer statistics. CA Cancer J. Clin. **60**(5), 277–300 (2010)
4. Iraqi Ministry of Health, Iraqi Cancer Board, Iraqi Cancer Register 2011, Registry Center Baghdad – Iraq (2014)
5. Gayathri, B.M., Sumathi, C.P., Santhanam, T.: Breast cancer diagnosis using machine learning algorithms-a survey. Int. J. Distrib. Parallel Syst. **4**(3), 105 (2013)
6. Khedr, A.E., Darwish, A., Ghalwash, A.Z., Osman, M.A.: Computer-aided early detection diagnosis system of breast cancer with fuzzy clustering means approach (2014)

7. Salahat, E.N., et al.: Methods and systems for processing MRI images to detect cancer, U.S. Patent Application No. 14/521,954 (2016)
8. Prannoy, G., Saravanakumar, K.: Breast cancer detection using image processing techniques. Int. Res. J. Comput. Sci. Technol. **10**, 391–399 (2017)
9. Gupta, M., et al.: Discrete wavelet transform-based color image watermarking using uncorrelated color space and artificial bee colony. Int. J. Comput. Intell. Syst. **8**(2), 364–380 (2015)
10. Zllner, F.G., Kyrre, E.E., Lothar, R.S.: SVM-based glioma grading: optimization by feature reduction analysis. Zeitschrift für medizinische Physik **22**(3), 205–214 (2012)
11. Haidekker, M.: Advanced Biomedical Image Analysis. Wiley, Hoboken (2010)
12. Kharat, K.D., Pradyumna, P.K., Nagori, M.B.: Brain tumor classification using neural network based methods. Int. J. Comput. Sci. Inform. **1**(4), 2231–5292 (2012)
13. El Abbadi, N.K., Faisal, Z.: Detection and analysis of skin cancer from skin lesions. Int. J. Appl. Eng. Res. **12**(19), 9046–9052 (2017)
14. Chen, D.-R., et al.: Use of the bootstrap technique with small training sets for computer-aided diagnosis in breast ultrasound. Ultrasound Med. Biol. **28**(7), 897–902 (2002)
15. Joo, S., et al.: Computer-aided diagnosis of solid breast nodules: use of an artificial neural network based on multiple sonographic features. IEEE Trans. Med. Imaging **23**(10), 1292–1300 (2004)
16. Keivanfard, F., et al.: Feature selection and classification of breast cancer on dynamic Magnetic Resonance Imaging by using artificial neural networks. In: 2010 17th Iranian Conference of Biomedical Engineering (ICBME). IEEE (2010)
17. Keyvanfard, F., Shoorehdeli, M.A., Teshnehlab, M.: Feature selection and classification of breast cancer on dynamic magnetic resonance imaging using ANN and SVM. Am. J. Biomed. Eng. **1**(1), 20–25 (2011)
18. Saini, S., Vijay, R.: Mammogram analysis using feed-forward back propagation and cascade-forward back propagation artificial neural network. In: 2015 Fifth International Conference on Communication Systems and Network Technologies. IEEE (2015)
19. Kaymak, S., Abdulkader, H., Uzun, D.: Breast cancer image classification using artificial neural networks. Procedia Comput. Sci. **120**, 126–131 (2017)
20. Arjao, T., et al.: Classification of breast cancer histology images using convolutional neural networks. PloS one **12**(6), e0177544 (2017)

# Author Index

Printed in the United States
By Bookmasters